AF874318

MINERALS, CRITICAL MINERALS, AND THE U.S. ECONOMY

Committee on Critical Mineral Impacts on the U.S. Economy

Committee on Earth Resources

Board on Earth Sciences and Resources

Division on Earth and Life Studies

NATIONAL RESEARCH COUNCIL

OF THE NATIONAL ACADEMIES

THE NATIONAL ACADEMIES PRESS

Washington, D.C.

www.nap.edu

THE NATIONAL ACADEMIES PRESS • 500 Fifth Street, N.W. • Washington, DC 20001

NOTICE: The project that is the subject of this report was approved by the Governing Board of the National Research Council, whose members are drawn from the councils of the National Academy of Sciences, the National Academy of Engineering, and the Institute of Medicine. The members of the committee responsible for the report were chosen for their special competences and with regard for appropriate balance.

This study was supported by the Department of the Interior / U.S. Geological Survey, under Award No. 06HQGR0204, and by the National Mining Association. The opinions, findings, and conclusions or recommendations contained in this document are those of the authors and do not necessarily reflect the views of the organizations or agencies that provided support for the project. Mention of trade names or commercial products does not constitute their endorsement by the U.S. Government.

International Standard Book Number-13: 978-0-309-11282-6
International Standard Book Number-10: 0-309-11282-6
Library of Congress Control Number: 2007943420

Additional copies of this report are available from the National Academies Press, 500 Fifth Street, N.W., Lockbox 285, Washington, DC 20055; (800) 624-6242 or (202) 334-3313 (in the Washington metropolitan area); Internet http://www.nap.edu

Cover: Conceptual illustration of the multifaceted nature of identifying and examining critical minerals. Nonfuel minerals are essential components of most of the products that are used every day including cellular telephones, computers, and electronics. Mineral availability to manufacture these products has to be evaluated in the context of a global mineral market and increasingly complex supply chains. Accurate and timely data on minerals are keys to inform public policy and avoid disruptive fluctuation in mineral supply. Design by Van Nguyen.

Copyright 2008 by the National Academy of Sciences. All rights reserved.

Printed in the United States of America.

THE NATIONAL ACADEMIES

Advisers to the Nation on Science, Engineering, and Medicine

The **National Academy of Sciences** is a private, nonprofit, self-perpetuating society of distinguished scholars engaged in scientific and engineering research, dedicated to the furtherance of science and technology and to their use for the general welfare. Upon the authority of the charter granted to it by the Congress in 1863, the Academy has a mandate that requires it to advise the federal government on scientific and technical matters. Dr. Ralph J. Cicerone is president of the National Academy of Sciences.

The **National Academy of Engineering** was established in 1964, under the charter of the National Academy of Sciences, as a parallel organization of outstanding engineers. It is autonomous in its administration and in the selection of its members, sharing with the National Academy of Sciences the responsibility for advising the federal government. The National Academy of Engineering also sponsors engineering programs aimed at meeting national needs, encourages education and research, and recognizes the superior achievements of engineers. Dr. Charles M. Vest is president of the National Academy of Engineering.

The **Institute of Medicine** was established in 1970 by the National Academy of Sciences to secure the services of eminent members of appropriate professions in the examination of policy matters pertaining to the health of the public. The Institute acts under the responsibility given to the National Academy of Sciences by its congressional charter to be an adviser to the federal government and, upon its own initiative, to identify issues of medical care, research, and education. Dr. Harvey V. Fineberg is president of the Institute of Medicine.

The **National Research Council** was organized by the National Academy of Sciences in 1916 to associate the broad community of science and technology with the Academy's purposes of furthering knowledge and advising the federal government. Functioning in accordance with general policies determined by the Academy, the Council has become the principal operating agency of both the National Academy of Sciences and the National Academy of Engineering in providing services to the government, the public, and the scientific and engineering communities. The Council is administered jointly by both Academies and the Institute of Medicine. Dr. Ralph J. Cicerone and Dr. Charles M. Vest are chair and vice chair, respectively, of the National Research Council.

www.national-academies.org

COMMITTEE ON CRITICAL MINERAL IMPACTS ON THE U.S. ECONOMY

RODERICK G. EGGERT, *Chair*, Colorado School of Mines, Golden
ANN S. CARPENTER, U.S. Gold Corporation, Reno, Nevada
STEPHEN W. FREIMAN, Freiman Consulting, Inc., Potomac, Maryland
THOMAS E. GRAEDEL, Yale University, New Haven, Connecticut
DREW A. MEYER, Vulcan Materials Company (retired), Birmingham, Alabama
TERENCE P. McNULTY, T.P. McNulty and Associates, Inc., Tucson, Arizona
BRIJ M. MOUDGIL, University of Florida, Gainesville
MARY M. POULTON, University of Arizona, Tucson
LEONARD J. SURGES, Natural Resources Canada, Ottawa, Ontario

National Research Council Staff

ELIZABETH A. EIDE, Study Director
NICHOLAS D. ROGERS, Research Associate

COMMITTEE ON EARTH RESOURCES

MURRAY W. HITZMAN, *Chair*, Colorado School of Mines, Golden
FRANCIS P. BURKE, CONSOL Energy, Inc. (retired), South Park, Pennsylvania
WILLIAM S. CONDIT, Bureau of Land Management (retired), Santa Fe, New Mexico
MICHAEL DOGGETT, Consultant, Vancouver, British Columbia, Canada
THOMAS V. FALKIE, Berwind Natural Resources Corporation (retired), Newtown Square, Pennsylvania
PATRICIA M. HALL, BP America Inc., Houston, Texas
DAVID D. LAURISKI, Safety Solutions International, LLC, Parker, Colorado
ANN S. MAEST, Stratus Consulting, Boulder, Colorado
LELAND L. MINK, U.S. Department of Energy Geothermal Program (retired), Worley, Idaho
REGINAL SPILLER, Frontera Resources Corporation, Houston, Texas
SAMUEL J. TRAINA, University of California, Merced
HAROLD J. VINEGAR, Shell Exploration and Production Company, Houston, Texas

National Research Council Staff

ELIZABETH A. EIDE, Senior Program Officer
NICHOLAS D. ROGERS, Research Associate

BOARD ON EARTH SCIENCES AND RESOURCES

GEORGE M. HORNBERGER, *Chair*, University of Virginia, Charlottesville
GREGORY B. BAECHER, University of Maryland, College Park
STEVEN R. BOHLEN, Joint Oceanographic Institutions, Washington, D.C.
KEITH C. CLARKE, University of California, Santa Barbara
DAVID J. COWEN, University of South Carolina, Columbia
WILLIAM E. DIETRICH, University of California, Berkeley
ROGER M. DOWNS, The Pennsylvania State University, University Park
JEFF DOZIER, University of California, Santa Barbara
KATHERINE H. FREEMAN, The Pennsylvania State University, University Park
RHEA L. GRAHAM, Pueblo of Sandia, Bernalillo, New Mexico
RUSSELL J. HEMLEY, Carnegie Institution of Washington, Washington, D.C.
MURRAY W. HITZMAN, Colorado School of Mines, Golden
V. RAMA MURTHY, University of Minnesota (retired), Minneapolis
CLAYTON R. NICHOLS, Idaho National Engineering and Environmental Laboratory (retired), Sandpoint
RAYMOND A. PRICE, Queen's University, Kingston, Ontario, Canada
BARBARA A. ROMANOWICZ, University of California, Berkeley
JOAQUIN RUIZ, University of Arizona, Tucson
MARK SCHAEFER, Foundation for Our Future, Reston, Virginia
WILLIAM W. SHILTS, Illinois State Geological Survey, Champaign
RUSSELL STANDS-OVER-BULL, Anadarko Petroleum Corporation, Billings, Montana
TERRY C. WALLACE, JR., Los Alamos National Laboratory, New Mexico
THOMAS J. WILBANKS, Oak Ridge National Laboratory, Oak Ridge, Tennessee

National Research Council Staff

ANTHONY R. DE SOUZA, Director
ELIZABETH A. EIDE, Senior Program Officer
DAVID A. FEARY, Senior Program Officer
ANNE M. LINN, Senior Program Officer
ANN G. FRAZIER, Program Officer
SAMMANTHA MAGSINO, Program Officer
CAETLIN M. OFIESH, Associate Program Officer
VERNA J. BOWEN, Administrative and Financial Associate
JENNIFER T. ESTEP, Administrative and Financial Associate
JARED P. ENO, Research Associate
NICHOLAS D. ROGERS, Research Associate
TONYA FONG YEE, Program Assistant

Preface

Earth resources—including those derived from the air, water, and land—are essential inputs into economic activity and fundamental determinants of our quality of life. The availability and quality of these resources and the adequacy of their supply, in both the short term and the long term, have been perennial national concerns over at least the last century. In the decades following World War II, for example, concerns centered around the adequacy of energy and mineral resources because of their significant use during the war and in postwar reconstruction. Indeed, throughout the entire latter half of the twentieth century, concerns have continually resurfaced about the reliability of supplies of key energy and mineral resources relevant for national security. In the 1970s, the nature of the concerns shifted to the short- and long-term reliability and availability of foreign sources of oil and other energy and nonfuel mineral resources such as bauxite and cobalt. A heightened public awareness of the importance of these resources, and of the long-term availability of all types of natural resources, became part of the national discussion. This awareness was one result of two decades of significant economic growth in North America and Europe, the beginning of the Japanese economic boom, and the emerging global interest in the effects of resource extraction and other human activities on environmental quality.

In the twenty-first century, the nature of the concerns over Earth resources has shifted once again. Energy and mineral commodity prices are relatively high for the first extended period since the 1970s, driven primarily by unexpectedly large demand growth in China, India, and other countries. At the same time, while the United States remains an important producer of energy and mineral resources, the extraction and production of these resources overall has shifted away from the United States toward other nations; U.S. import dependence for many commodities has increased and has raised concerns about reliability of the foreign supply.

While the issue of foreign dependence on oil and gas is a dominant theme in the media and in discussions regarding national energy policy, the advent of the Information Age has demanded an ever-wider range of metallic and nonmetallic minerals to perform essential functions in new products such as computers, cellular telephones, and transportation equipment. Although no crisis of national proportions—or proportions similar to the oil embargo of 1973, for example—has occurred with regard to nonfuel mineral availability for the United States, technical advances and the globalization of the mineral market raise the question, will the necessary nonfuel mineral resources be available in time and at acceptable costs to meet burgeoning demand for these and other emerging products and technologies? Growing recognition that primary nonfuel mineral producers must give greater priority to the social and environmental consequences of mineral extraction for local communities and the natural environment, and that secondary resources from recycling are underexploited, add further to the set of issues being considered in determining national approaches for responsible and effective acquisition and use of nonfuel mineral resources.

Within that broad context, this study was motivated by concerns expressed to members of the Committee on Earth Resources of the National Research Council (NRC) by the nonfuel minerals community regarding the nature and adequacy of federal support both to understand the nonfuel minerals that are important to the nation's economy and functions, and to collect nonfuel mineral data appropriate for making informed policy decisions that help avoid restrictions in nonfuel mineral supply. This study was intended to address nonfuel mineral issues in advance of a national crisis with the idea that it is potentially prudent and cost-effective to determine policy and appropriate action before any such crisis occurs. For these reasons, the Committee on Critical Mineral Impacts on the U.S. Economy, appointed by the NRC, was asked to identify and review nonfuel minerals that are "critical" for domestic industry and emerging technologies; assess their global trends in sources and production; examine potential constraints on their availability; identify impacts of restrictions in their supply on the domestic economy; and describe and evaluate current and future nonfuel mineral information, databases, and research that could enhance the understanding of mineral criticality in a global context. The committee comprises individuals with expertise in nonfuel mineral exploration and ore deposits, mineral economics, metallurgy, statistics, federal and international standards, regulatory policy, recycling, industrial materials and manufacturing, and mineral processing and engineering, including nanotechnology. More committee information is available in Appendix A.

The committee found the task of identifying critical minerals to be an exciting and multifaceted challenge and worked diligently to find a framework that would capture the dynamic nature of critical minerals. Although it is not a panacea, the "criticality matrix" described in this report became the organizing framework around which we developed our assessment. This matrix serves as a conceptual lens through which to view and assess mineral criticality. During the course of this study, the committee was convinced of the importance of the availability of consistent, unbiased data and analysis on a complete suite of nonfuel minerals that were, are now, or may become critical. Informed decisions on nonfuel minerals depend on access to these types of data.

All members of the committee provided key insights and took part in drafting this report. The committee was ably assisted by, and in fact could not have functioned effectively without, the NRC staff members assigned to this study, Senior Program Officer Elizabeth A. Eide and Research Associate Nicholas D. Rogers. We thank them heartily. We also thank the reviewers for their criticisms and constructive suggestions for the report.

Nonfuel minerals will continue to be important to U.S. consumers and the economy. We hope our assessment will be helpful to decision makers and the general public as they evaluate the role of minerals in the economy and will inform public policies regarding mineral information and minerals more generally.

Roderick G. Eggert
Chair
August 2007

Acknowledgments

In addition to its own expertise, the committee relied on input from numerous external professionals with extensive experience in nonfuel minerals and materials research, industrial applications for minerals, mineral exploration, and government and environmental policy from federal agencies, academic institutions, and the private sector. These individuals provided testimony on which nonfuel minerals they found critical to their own work, which minerals might become critical in the future, and which data were required, collected, and accessible to aid them in making better or more efficient decisions with regard to access to these minerals. This information was extremely important to the committee in formulating its report, and we would like to express our appreciation to the many highly qualified individuals who provided testimony, data, and advice during the course of the study; in particular, the committee would like to thank Mark Barton, John Benner, David Cammarota, Catherine Cochran, Rick Deery, John DeYoung, Mark Ellis, Jason Goulden, Rich Heig, Ivan Herring, Anthony Hodge, Jeremiah Johnson, Kate Johnson, Phil Jones, Michael Kaas, Pat Leahy, Marc LeVier, James Marder, Dave Menzie, Glenn Miller, John Morgan, Jr., Shinsuke Murakami, Lauren Pagel, Gina Pearson, Carol Raulston, Joanne Shore, Lew Sloter, and Larry Stevens.

This report has been reviewed in draft form by individuals chosen for their diverse perspectives and technical expertise, in accordance with procedures approved by the National Research Council's Report Review Committee. The purpose of this independent review is to provide candid and critical comments that will assist the institution in making its published report as sound as possible and to ensure that the report meets institutional standards for objectivity, evidence, and responsiveness to the study charge. The review comments and draft manuscript remain confidential to protect the integrity of the deliberative process. We wish to thank the following individuals for their participation in the review of this report:

William X. Chavez, Jr., New Mexico Institute of Mining and Technology, Socorro
Harry M. Conger, Homestake Mining Company, Palo Alto, California
Rhea L. Graham, Pueblo of Sandia, Bernalillo, New Mexico
Glenn C. Miller, University of Nevada, Reno
Robert Reiley, U.S. Department of Commerce (retired), Reston, Virginia
Mark C. Roberts, Michigan Technological University, Houghton
Steven D. Scott, University of Toronto, Ontario, Canada
Spencer R. Titley, University of Arizona, Tucson
Kathleen Walsh, U.S. Naval War College, Newport, Rhode Island

Although the reviewers listed above have provided many constructive comments and suggestions, they were not asked to endorse the conclusions or recommendations nor did they see the final draft of the report before its release. The review of this report was overseen by Jean-Michel M. Rendu, mining consultant, and William G. Agnew, General Motors Corporation (retired). Appointed by the National Research Council, they were responsible for making certain that an independent examination of this report was carried out in accordance with institutional procedures and that all review comments were carefully considered. Responsibility for the final content of this report rests entirely with the authoring committee and the institution.

Contents

Summary

INTRODUCTION

The unique properties of nonfuel minerals, mineral products, metals, and alloys contribute to the provision of food, shelter, infrastructure, transportation, communications, health care, and defense. Every year more than 25,000 pounds (11.3 metric tons) of new nonfuel minerals must be provided for each person in the United States to make the items that we use every day. As our ability has advanced to produce new materials and to characterize, predict, and exploit the chemical and physical properties of nonfuel minerals, it has become possible to develop new applications that improve the technical performance, durability, and reliability of products; to deliver greater value to businesses and consumers; and to reduce environmental burdens. In the modern age, developments in materials science and engineering, mineral exploration, and processing continue to enable and support new technologies; the existence or function of common items such as cellular telephones, computers, automobiles, toothpaste, paint, or a stable electrical supply could not be possible without nonfuel minerals or mineral products and related materials. Minerals are thus fundamental inputs to the domestic economy and daily life at scales ranging from the individual consumer to entire manufacturing and engineering sectors.

Many existing and emerging technologies require nonfuel minerals that are not available in the United States. The global nature of the nonfuel mineral market has been made very evident in recent years as many emerging economies have become significant as both producers and consumers of various raw mineral products, in some cases competing for mineral feedstock directly with U.S. producers, manufacturers, and users. While foreign competition for minerals is one aspect in a range of factors affecting the supply of minerals to the U.S. economy, a high degree of import dependence for certain minerals is not in itself a reason for concern. However, import dependence can expose a range of U.S. industries to political, economic, and other risks that vary according to the particular situation. Informed planning to maintain and enhance domestic economic growth requires knowledge of potential restrictions in the supply of nonfuel minerals, as well as the strategies to mitigate the effects of those restrictions.

This study was an outgrowth of discussions during the past several years with the Committee on Earth Resources of the National Research Council (NRC) on the topic of nonfuel minerals, their availability and use in domestic applications, and their national importance in a global mineral market. The committee was concerned that the impacts of potential restrictions on the supply of nonfuel minerals to different sectors of the U.S. economy were not adequately articulated in the national discussion of natural resource use. In addition, federal responsibilities to acquire and disseminate information and conduct research on critical nonfuel minerals were not well defined in a framework that also accounts for the complete, global mineral cycle, from exploration to recycling. Positive response to the committee's formulation of a study prospectus by several federal agencies and professional organizations, including two of the study's eventual sponsors, the U.S. Geological Survey and the National Mining Association, encouraged the NRC to establish the Committee on Critical Mineral Impacts on the U.S. Economy (Appendix A); this report is the committee's response to the study's statement of task (Box S.1).

This report investigates and highlights the importance of nonfuel minerals in modern U.S. society; which minerals might be termed "critical" and why; the extent to which the availability of these minerals is subject to restriction in the short to the long term; and when considering mineral criticality, which data, information, and research are needed to aid decision makers in taking appropriate steps to mitigate restrictions in the nonfuel mineral supply. The audience for the study includes not only federal agencies, industry, and research organizations, but necessarily also the general public and decision makers.

Chapter 1 establishes the basic methodology used by the committee to determine mineral criticality in the framework of a "criticality matrix"

BOX S.1
Statement of Task

Understanding the likelihood of disruptive fluctuation in the supply of critical minerals and mineral products for domestic applications, and making decisions about policies to reduce such disruptions, requires thorough understanding of national and international mineral sources, mineral production technology, the key uses of minerals and mineral products in the United States economy, and potential impediments to the mineral supply.

This study will:

1. Identify the critical minerals and mineral products that are essential for industry and emerging technologies in the domestic economy (addressed in Chapters 1-3 and in culminating discussion in Chapter 4);
2. Assess the trends in sources and production status of these critical minerals and mineral products worldwide (addressed in Chapters 3 and 4);
3. Examine the actual or potential constraints, including but not limited to geologic, technological, economic, and political issues, on the availability of these minerals and mineral products for domestic applications (addressed in Chapters 3 and 4);
4. Identify the impacts of disruptions in supply of critical minerals and mineral products on the domestic workforce and economy (addressed in Chapter 2);
5. Describe and evaluate the current mineral and mineral product databases and other sources of mineral information available for decision making on mineral policy issues (addressed in Chapter 5); and
6. Identify types of information and possible research initiatives that will enhance understanding of critical minerals and mineral products in a global context (addressed in Chapter 5).

(Figure S.1) and establishes the concept of the "material flow" or "life-cycle" approach to assessing minerals and their criticality. Chapter 2 examines the vertical axis of the criticality matrix—the *importance of minerals in use*—through examples that demonstrate specific applications of minerals and materials in some important U.S. industrial sectors and the significance of the degree of mineral *substitutability* in these applications. Chapter 3 examines the horizontal axis of the criticality matrix—the *availability* and

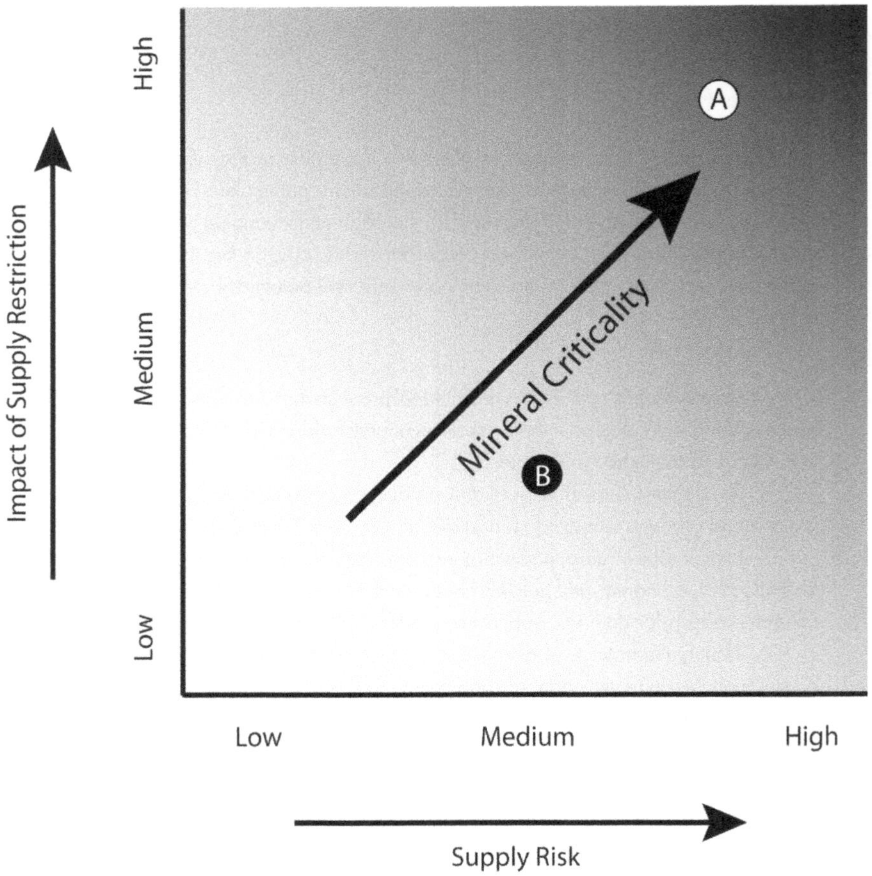

FIGURE S.1 The criticality matrix as established in this report allows evaluation of the "criticality" of a given mineral. A specific mineral or mineral product can be placed on this figure after assessing the impact of restriction on the mineral's supply should it occur (vertical axis) and the likelihood of a supply restriction (horizontal axis). The degree of criticality increases as one moves from the lower-left to the upper-right corner of the figure. In this example, mineral A is more critical than mineral B. More quantitative descriptions of the parameters used to evaluate mineral supply restrictions and their impacts are presented in this report in terms of composite scores for 11 minerals assessed in this study.

reliability of the mineral supply (supply risk)—to more completely describe the numerous factors that can affect mineral supply from short- through long-term periods. Factors affecting the availability of primary (virgin ore), secondary (e.g., scrap or recycled products), and tertiary (imported goods presently in service) mineral sources are also examined. Chapter 4 demonstrates the application of the criticality matrix methodology to evaluate mineral criticality by examining 11 mineral candidates for criticality. The minerals and their applications cross many more industry sectors than the four examined in detail in Chapter 2 and serve to underscore the ubiquitous applications for minerals in everyday life. The committee specifically examined copper, gallium, indium, lithium, manganese, niobium, platinum group metals (PGMs: platinum, palladium, rhodium, iridium, osmium, ruthenium), rare earth elements (REs: lanthanum, cerium, praseodymium, neodymium, promethium, samarium, europium, gadolinium, terbium, dysprosium, holmium, erbium, thulium, ytterbium, lutetium), tantalum, titanium, and vanadium to determine their criticality. The committee did not have the time or resources to evaluate all potentially critical minerals; the process used to select and examine these minerals is explained in more detail below. Chapter 5 presents an overview of the federal data, information, and research appropriate for making informed decisions about minerals in general and critical minerals in particular. The main conclusions and recommendations of the study are presented in Chapter 6.

The committee was fortunate in that a complementary study was conducted concurrently by the National Materials Advisory Board of the NRC in which an ad hoc committee assessed the need for a National Defense Stockpile. Therefore, the present committee has included in this report only limited discussion on mineral needs and issues specific to the defense sector and refers the reader to this other NRC report for detailed information.

THE METHODOLOGY: A MINERAL CRITICALITY MATRIX

As the main part of its assessment, the committee developed a methodology—a criticality matrix—to use in assessing a nonfuel mineral's degree of criticality (Figure S.1). The methodology provides a framework for federal agencies, decision makers, the private sector, and any users interested in minerals to make assessments about their own critical minerals and, on that basis, to determine which data, information, and research might be necessary to mitigate potential restrictions in the supply of that mineral for an existing or future use. To be critical, a mineral must be both essential in use (represented on the vertical axis of the matrix) and subject to supply restriction (the horizontal axis of the matrix). Keys to understanding how essential a mineral is in a particular application are its chemical and physical properties. The corollary is that some minerals may be of more concern than others—or have greater *importance in use*—in the sense that few if any substitutes can provide similar functionality at comparable costs. A key to understanding supply restrictions is recognizing that availability—and thus, restriction on availability—depends on the time scale of interest. Fundamentally and over the long term (more than about 10 years), availability is a function of geologic, technical, environmental and social, political, and economic factors. In the short to medium term (a few months to a few years, but less than a decade), availability and reliability of supply can be assessed using a variety of market-specific factors such as worldwide mineral reserve-to-production ratios, world by-product production, U.S. secondary production (through scrap and recycling), import dependence, and the degree to which production is concentrated in a small number of companies or countries. This combination of *importance in use* and *availability or supply risk*—specifically, the potential that an important nonfuel mineral may be subject to supply restrictions—defines a mineral's criticality for any specified time scale.

IMPORTANCE OF MINERALS IN USE

In 2006, the overall value added to the annual gross domestic product by industries that consume processed nonfuel minerals was estimated to be in excess of $2.1 trillion. Mineral industry-related employment, broadly defined, in 2006 was close to 1.5 million jobs. These figures indicate that overall mineral use and associated employment are important to the U.S.

economy. On the other hand, as discussed later, not all mineral use is subject to supply disruption; thus, these figures ought not to be interpreted as indicators of how much of the U.S. economy is at risk should the supply of a specific mineral be restricted.

Understanding the importance of nonfuel minerals in the products from different sectors of the U.S. economy forms the basis for the vertical axis of the committee's criticality matrix, and minerals have varying levels of importance on this axis as a result of the demand for them in a particular end use. "Importance in use" carries with it the concept that some minerals will be more fundamental for specific uses than others, depending on the mineral's chemical and physical properties. Again, by focusing on a mineral's properties, the role of *substitutability* is a key factor in a mineral's importance: a mineral for which substitutes are easily found is going to be slightly less important than one for which substitutes that provide the same properties, at comparable costs, cannot be found in the short term.

Four important industry sectors in the United States—the automotive, aerospace, electronic, and energy sectors—illustrate the use of different minerals in various common applications and highlight the specificity of particular chemical and physical mineral properties for an application. These four sectors were chosen as useful examples for purposes of discussion in this report of mineral criticality, but similar arguments could be made for other important sectors such as health care, construction, utilities, or the transportation infrastructure. All sectors of the economy rely on the services provided by minerals, and the committee was limited by its time and resources to selecting a few industry examples. In determining criticality of specific minerals later in the report, however, the committee took a broader view and included all industry applications, across all industry sectors. In addition to the relationship between mineral properties and mineral substitutability, examination of industrial sectors also lends itself to discussion of the concept that mineral criticality is dynamic: technology advancements, the popularity of given products, the discovery of health or environmental issues related to mineral use, or the rise of regulatory, tariff, or trade issues can all change the level of demand for one or another mineral through time.

An example from the automotive industry illustrates this point. Automobiles manufactured at the start of the twentieth century were composed of about five materials: wood, rubber, steel, glass, and brass. Today, a typical automobile may contain up to 39 different nonfuel minerals in various components, in addition to rubber, plastic, and other organically based materials. Copper, for example, has become the preferred metal for electrical wiring in today's automobiles. About 50 pounds of copper is used in an average automobile today. The new technology of hybrid electric cars requires greater amounts of copper—about 75 pounds in total, by some estimates. PGMs and REs are other families of minerals fundamental to the construction and function of automobile catalytic converters. At present, no viable substitutes exist for these minerals in this application, resulting essentially in a "no-build" situation for catalytic converters should PGM or RE supply be restricted. The report also discusses other examples of mineral applications for the propulsion system and structural frames of airplanes (e.g., titanium), cellular telephones (e.g., tantalum), liquid crystal displays (e.g., indium), computer chips (a broad mineral suite), photovoltaic cells (e.g., silicon, gallium, cadmium, selenium, tellurium, indium), and rechargeable batteries (e.g., lithium, REs, nickel). These examples support the finding that mineral importance changes through time and is largely a function of the properties of minerals and mineral substitutability in a given application.

MINERAL AVAILABILITY OR RISK TO SUPPLY

In evaluating the horizontal axis of the criticality matrix, five dimensions of primary, long-term nonfuel mineral availability were identified: geologic (whether the mineral resource exists), technical (whether we extract and process it), environmental and social (whether we can produce it in environmentally and socially accepted ways), political (whether governments influence availability through their policies and actions), and economic (whether we can produce it at a cost that users are willing and able to pay). With the exception of geologic availability, the same factors apply to the secondary availability of a mineral. Instead of being dependent on its geologic occurrence, the magnitude of the secondary resource depends on past inflows and outflows from the stock of materials available for recycling, including material discarded in landfills.

In the short and medium term, significant restrictions of supply may exist, leading either to physical unavailability of a nonfuel mineral or, more likely, to higher prices. First, *demand may increase significantly and unexpectedly*, and if production already is occurring at close to capacity, either a mineral will become physically unavailable or its price will rise significantly. Second, *relatively thin (or small) markets* may make it difficult to increase production quickly if demand increases significantly. Third, supply may be prone to restriction if *production is concentrated* in a small number of mines, a small number of companies, or a small number of producing countries. Fourth, minerals whose supply consists significantly of *by-product production* may be fragile or risky because the availability of a by-product is determined largely by the availability of the main product (e.g., gallium as a by-product of bauxite mining). Finally, minerals for which there is not significant *recovery of material from old scrap* may be more prone to supply risk than otherwise. Other possible indicators of supply risk, which are commonly cited and possibly useful but only if interpreted with care, are *import dependence*, the *reserve-to-production ratio*, and the ratio of *reserve base-to-production*. These factors can easily be misinterpreted. As outlined earlier in this summary, for example, high measured import reliance does not necessarily imply that supply is at risk. In fact, in several situations, high measured import reliance may be no more risky than domestic supply. The committee found that a balanced interpretation of all of these factors in terms of examination of supply risk is highly dependent on good domestic and global data on nonfuel minerals and mineral markets and comprehensive and reliable analysis of such data.

CRITICAL MINERAL CANDIDATES: APPLICATION OF THE MATRIX

The committee used the established parameters regarding a mineral's *importance in use* and *availability* (supply risk) to apply the criticality matrix to 11 minerals or mineral groups: copper, gallium, indium, lithium, manganese, niobium, PGMs, REs, tantalum, titanium, and vanadium. Because the committee did not have the time or resources to evaluate all potentially critical minerals, these 11 were selected on the basis of two considerations. First, the set of minerals the committee examined had to illustrate the range of circumstances that the matrix methodology accommodates and considers. For example, in the selection of the minerals examined in this report, the committee considered minerals used in large quantities throughout the economy in traditional applications and others used in limited quantities in a small number of (often emerging) applications, minerals produced largely as by-products, and other minerals for which recycling of scrap is an important source of supply. Second, the set of minerals had to consist of those that, in the professional judgment of committee members, would likely be included in a more comprehensive assessment of all potentially critical minerals. The committee used a combination of quantitative measures and expert (qualitative) judgment in implementing the matrix methodology.

Recognizing that restriction in an individual mineral's supply will not have the same macroeconomic impact on the nation as a restriction in the supply of oil, the committee evaluated the criticality of each of these minerals on the basis of whether or not a particular industry sector, or the manufacture of one or more fairly ubiquitous consumer products, would be adversely affected should a restriction on the supply of that mineral occur. The committee acknowledges that some minerals not considered specifically in this report could be soundly argued to be critical to a particular industry, individual, state, community, or federal agency and encourages application of the matrix or similar methodology for these specific needs. Of the 11 minerals examined, PGMs, REs, indium, manganese, and niobium were determined to be critical in the sense that the importance of their applications (in automotive catalytic converters, industrial chemical production, electronics, batteries, liquid crystal displays, or hardeners or strengtheners in steel and iron alloys), the difficulty in finding appropriate mineral substitutes for these applications, and the risk to their supply for any one of a number of reasons were high enough to place these minerals in or near the critical zone on the criticality matrix. Although important applications exist for the other minerals examined by the committee (copper, gallium, lithium, tantalum, titanium, and vanadium), ready substitutes or low risk to their supplies indicate that these minerals are not potentially prone to restriction at present. The committee notes that it did not speculate on the potential for new, or frontier, applications to drive new demand for these or other minerals in the future, underscoring the committee's emphasis on the need for federal collection, analysis, and dissemination of current and consistent data and information on all minerals.

Finally, the committee's application of the matrix methodology—and the particular example of copper—highlights the distinction between minerals that are *essential* to the economy in certain applications but are yet *not critical*, at least at present, because the risk of supply restriction is low. Other minerals likely to fall in this category of essential, but not critical, include bauxite (the mineral raw material for aluminum), iron ore, and construction aggregates.

INFORMATION AND RESEARCH ON MINERALS

Although a wide range of government and nongovernmental, international, and domestic organizations collects and disseminates information and databases relevant for decision making on nonfuel mineral policy issues, the committee found that decision makers in both the public and the private sectors desire continuous, unbiased, and thorough mineral information provided through a federally funded system of information collection and dissemination. Historically, nonfuel mineral data collection in the United States has been a recognized part of national policy since at least World War I, with a foundation in the importance of minerals to the national economy and national security and an emphasis on the importance of good statistical data collection to inform policy decisions. Of the more than 70 federal agencies or programs that receive funding to carry out statistical activities of all types, only 13 are considered "principal" federal statistical agencies. The committee finds it significant that none of the 13 principal statistical agencies collects and publishes annual data on nonfuel minerals and their availability. These principal statistical agencies obtain nearly half of the funding allocated to agencies for statistical data collection, and since their main missions are oriented toward the collection and dissemination of specific types of data, they have a certain degree of autonomy, focus, and ability to maintain the levels of data collection and analysis necessary to support those missions. Agencies that collect data as only one part of their mission may find data collection to be diluted or made subordinate to the overall mission of the agency or department of which they are a part, particularly in situations of constrained resources.

In its examination of federal and nongovernmental sources of information on minerals, the committee concurred with the consensus view of private, academic, and federal professionals that the U.S. Geological Survey (USGS) Minerals Information Team is the most comprehensive, responsible, and responsive source of information on nonfuel minerals both domestically and internationally but that the quantity and depth of its data and analysis have fallen in recent years, due at least in part to reduced or static budgets and concomitant reductions in staff and data coverage. Because the effectiveness of a government agency or program is dependent on the agency's or program's autonomy, its level of resources, and its authority to enforce data collection, federal information gathering for nonfuel minerals as presently configured does not have sufficient authority and autonomy to appropriately carry out its data collection, dissemination, and analysis.

In addition to the types of data already appropriately collected by the Minerals Information Team and the resource assessments and research on mineral exploration, production, consumption, and environmental impacts conducted by the USGS Mineral Resources Program under which the Minerals Information Team is administered, the committee is supportive of the incorporation of "critical minerals" as a specific part of analysis and data collection in the context of the complete mineral life cycle. The committee finds domestic and international data collection in the following areas to be useful in this regard: recycling or scrap generation and inventories of old scrap; in-use stocks; reserves and resources; downstream uses; subeconomic resources; and material flows. There currently is no common federal source of information that supplies all of the above data on at least an annual basis for all minerals.

In a number of areas relevant to critical minerals, the committee found a paucity of information. The committee relates these deficiencies to an inappropriately low level of support for research related to resource availability and resource technology. In particular, the following research topics are important if critical minerals are to be reliably identified in the future, if the sources of those minerals (from both virgin and recyclable stocks) are to be better quantified, and if the technology for their extraction and processing is to be substantially enhanced. Areas considered important in this regard include: theoretical geochemical research; extraction and processing and waste disposal technology to improve energy efficiency, to decrease water use, and to enhance material separation; remanufacturing and recycling technology; and the characterization of stocks and flows of materials (especially import and export) as components of products and the losses upon product discard. Aside from nonfuel mineral data collection and analysis, the committee supports federal roles in facilitating and enabling technology transitions and monitoring of markets for new technological applications that employ minerals; the fact that many materials in new applications have come about through government involvement in research and development to achieve higher performance provides validation for this approach.

Finally, the committee found that well-educated resource professionals are essential for fostering the innovation that is necessary to ensure resource availability at acceptable costs with minimal environmental damage. The infrastructure for adequate training of professionals to service the mineral and materials sectors has declined substantially over the past few decades in almost all industrialized countries. The current pipeline of training in the United States does not have enough students to fill the present or anticipated future needs of the country in terms of mineral resource capabilities in the private sector, the federal government, academic institutions, particularly if critical minerals are to be part of the government's mineral data collection, analysis, and dissemination program. While market responses may eventually cover some of the apparent gap between the short-term demand for workers and the supply of new hires, the time lag of market responses, the very large number of anticipated workforce openings, and the need for technology innovation entail larger commitments than the market alone is able to address and suggest the need for government engagement in the matter of professional training.

CONCLUSIONS

Based upon its analysis, the committee draws the following main conclusions:

- **All minerals and mineral products could be or could become critical to some degree, depending on their importance and availability.**
- **From the federal perspective, a critical mineral is one that is both essential in use *and* subject to the risk of supply restriction.**
- **The criticality of a specific mineral can and likely will change as production technologies evolve and new products are developed.**
- **The greater the difficulty, expense, or time it takes for material substitution to occur, the more critical a mineral is to a specific application or product—or analogously, the greater is the impact of a mineral supply restriction.**
- **The criticality matrix methodology is a useful conceptual framework for evaluating a mineral's criticality in a balanced manner in a variety of circumstances that will be useful for decision makers in the public and private sectors.** A more nuanced and quantitative version of the matrix could be established and used as part of the federal program for collection, analysis, and dissemination of data on minerals.
- In employing the methodology, **it is important to distinguish among three time or adjustment periods: the short term, the medium term, and the long term.**
- **In the short and medium terms, significant restrictions of mineral supply may be due to (1) significant increase in demand, (2) thin markets, (3) concentration of production, (4) production**

predominantly as a by-product, (5) lack of available old scrap for recycling or of the infrastructure required for recycling.

- **Over the longer term, the availability of minerals and mineral products is largely a function of investment and the various factors that influence the level of investment and its geographic allocation and success. The long-term availability of minerals and mineral products also requires continued investment in mineral education and research.**
- As an indicator of vulnerable supply, **import dependence by itself is not a useful indicator of risk.** Rather, to be vulnerable to supply restriction, some other factor must be present that makes imports vulnerable to disruption—for example, supply is concentrated in one or a small number of exporting nations with high political risk or in a nation with such significant growth in internal demand that exported minerals may be redirected toward internal, domestic use.
- Of the 11 minerals or mineral families the committee examined, **those that exhibit the highest degree of criticality at present are indium, manganese, niobium, PGMs, and REs.**
- **Decision makers in both the public and the private sectors need continuous, unbiased, and thorough mineral information provided through a federally funded system of information collection and dissemination.**
- **The effectiveness of a government agency or program is dependent on the agency's or program's autonomy, its level of resources, and its authority to enforce data collection. Federal information gathering for minerals at present does not have sufficient authority and autonomy to appropriately carry out its data collection, dissemination, and analysis. In particular, the committee concludes that USGS Minerals Information Team activities are less robust than they might be, in part because it does not have the status or resources to function as a "principal" statistical agency.**

- **More complete information needs to be collected, and more research needs to be conducted, on the full mineral life cycle.**

RECOMMENDATIONS

Recognizing the dynamic nature of mineral supply and demand and of criticality, and in light of the above conclusions, the committee makes the following recommendations:

1. The federal government should enhance the types of data and information it collects, disseminates, and analyzes on minerals and mineral products, especially as these data and information relate to minerals and mineral products that are or may become critical.

In particular, more attention needs to be given to those areas of the mineral life cycle that are underrepresented in current activities including reserves and subeconomic resources; by-product and coproduct primary production; stocks and flows of secondary material available for recycling; in-use stocks; material flows; international trade, especially of metals and mineral products embodied in imported and exported products; and related information deemed appropriate and necessary. Enhanced mineral analysis should include periodic assessment of mineral criticality using the committee's matrix methodology or some other suitable method.

2. The federal government should continue to carry out the necessary function of collecting, disseminating, and analyzing mineral data and information. The USGS Minerals Information Team, or whatever federal unit might later be assigned these responsibilities, should have greater authority and autonomy than at present. It also should have sufficient resources to carry out its mandate, which would be broader than the Minerals Information Team's current mandate if the committee's recommendations are adopted. It should establish formal mechanisms for communicating with users, government and nongovernmental organizations or institutes, and the private sector on the types and quality of data and information it collects, disseminates, and analyzes. It should

be organized to have the flexibility to collect, disseminate, and analyze additional, nonbasic data and information, in consultation with users, as specific minerals and mineral products become relatively more critical over time (and vice versa).

The Energy Information Administration provides a potential organizational model for such an agency or administrative unit. The federal government should consider whether a comparable mineral information administration would have status as a principal statistical agency and, if not, what other procedures should be investigated and implemented to give an agency with the mandate to collect mineral data and information greater autonomy and authority, as well as sufficient resources, to carry out its mandate. In the globalized mineral market, it is essential that the United States has a central authority through which to conduct outreach and exchange programs on minerals data with international counterparts and to collect and harmonize data from international sources. Combined U.S. government and foreign government efforts are likely to provide the most accurate, uniform, and complete data sets of this information over time and thereby provide adequate information to all communities concerned about future global mineral or material supply and demand trends.

3. Federal agencies, including the National Science Foundation, Department of the Interior (including the USGS), Department of Defense, Department of Energy, and Department of Commerce, should develop and fund activities, including basic science and policy research, to encourage U.S. innovation in the areas of critical minerals and materials and to enhance understanding of global mineral availability and use.

Without renewed federal commitment to innovative research and education on minerals, it is doubtful whether the activities recommended in this report regarding information about minerals will be sufficient for the nation to successfully anticipate and respond to possible short- to long-term restrictions in mineral markets.

CHAPTER 1

Critical Minerals

INTRODUCTION

Archaeologists and historians describe early civilizations and periods of human history using terms such as the Stone Age, the Copper Age, the Bronze Age, and the Iron Age. Such descriptions reflect the fundamental importance of nonfuel minerals, metals, and materials technology and applications. Early civilizations were built to a significant degree using the seven metals of antiquity (in order of discovery): gold (6000 BC), copper (4200 BC), silver (4000 BC), lead (3500 BC), tin (1750 BC), iron (1500 BC), and mercury (750 BC). Each discovery led to a range of innovations and applications that provided a marked advantage until such time as it was adopted by competing civilizations or overtaken by other innovations. Advances were not limited to military technology but extended to agricultural implements, food storage and preparation, therapeutic and cosmetic applications, and many other aspects of daily life and culture. The much later discovery of arsenic, antimony, zinc, and bismuth in the thirteenth and fourteenth centuries was followed by platinum in the sixteenth century and another 12 metals or metalloids in the eighteenth century, bringing the total number of known metals and metalloids to 24. Most known metals and metalloids were discovered only in the past two centuries.

As our ability has advanced to produce new materials and to characterize, predict, and exploit the chemical and physical properties of minerals, it has become possible to develop new applications that improve the technical performance, durability, and reliability of products; deliver greater value to businesses and consumers; and reduce environmental burdens. In the Information Age, developments in materials science and engineering, mineral exploration, and processing continue to enable and support the development of new technologies. The unique properties of nonfuel minerals, mineral products, metals, and alloys contribute to provision of food, shelter, infrastructure, transportation, communications, health care, and defense. The cellular telephone is one familiar example that illustrates the dependence of a globally important technology on minerals, their chemical and physical properties, and the materials created with them (Box 1.1).

Products depend on essential nonfuel minerals and mineral products the supply of which is sometimes subject to disruption or restriction. In the short or medium term (a few months to a decade), the balance between demand and supply often is fragile and prices may thus be volatile. Over the longer term (more than about 10 years), the availability of nonfuel minerals and mineral products depends heavily on investments in people and technology. Insufficient investment today can lead to restrictions on availability in the future.

This study was undertaken to investigate and highlight both the importance of nonfuel minerals and mineral products in modern U.S. society and the extent to which the availability of these minerals and mineral products is subject to restriction in both the short to medium terms and the long term.

BACKGROUND OF STUDY AND COMMITTEE CHARGE

This study was an outgrowth of meeting discussions and professional exchanges during the past several years conducted by the Committee on Earth Resources (CER) of the National Research Council (NRC) on the topic of nonfuel minerals, their availability and use in domestic applications, and their continued national importance in a global mineral market. The committee was concerned that the impacts of potential restrictions on the supply of nonfuel minerals to different sectors of the U.S. economy were not adequately articulated in the national discussion of natural resource use, and that federal responsibilities to acquire and disseminate information and conduct research on nonfuel minerals were not well defined in a global framework that also accounts for the complete mineral cycle from exploration to recycling (NRC, 2004a, b). Aware of the numerous NRC studies on the topics of nonfuel minerals, federal minerals policy, and federal programs tasked with mineral research and information, the committee suggested that recognition of those minerals that could be considered pivotal, or "critical," for a particular industrial, civilian, or military sector is an important aspect of the nonfuel mineral discussion that had not yet been addressed in an independent NRC report. The committee drafted a prospectus for a study designed to inform public policy on critical mineral impacts on the U.S. economy.

Numerous federal agencies and professional organizations were asked by CER to comment on the committee's study formulation. The U.S. Geological Survey (USGS) and the National Mining Association deemed the issue sufficiently important to bring forward to public discussion in the form of an NRC study. The NRC thus established the Committee on Critical Mineral Impacts on the U.S. Economy to address the issues outlined in the study's statement of task (Box 1.2). The committee consists of nine experts from academia, industry, the federal sector, and Natural Resources Canada. These individuals contributed their professional expertise in areas of mineral exploration and ore deposits, mineral economics, metallurgy, statistics, federal and international standards, regulatory policy, recycling, industrial materials and manufacturing, and mineral processing and engineering, including nanotechnology (Appendix A).

This report constitutes the committee's response to the study charge. This first chapter reviews some of the important issues the study committee has extracted from previous NRC reports on the topic of nonfuel minerals and the mineral life cycle, the committee's interpretation of the term "critical mineral," and the framework in which the committee found it most useful to determine a mineral's criticality.

PREVIOUS AND ONGOING WORK

A number of NRC reports published during the last two decades have analyzed or evaluated a variety of national nonfuel mineral issues. The

BOX 1.1
Cellular Phones, Minerals, and Technology

Cellular phones have become omnipresent in today's society; in 2006, worldwide mobile phone sales exceeded one billion units, an increase of about 21 percent over sales in 2005 (*http://www.gartner.com/it/page.jsp?id=501734*; accessed September 5, 2007). What is not widely recognized is the dependence of cell phone performance, and therefore the communications system, on a wide variety of minerals, many of which can be scarce or expensive to process.

The technological barrier to cellular communication was overcome only in the 1970s with the discovery of barium titanate ceramics. These ceramics possess the requisite dielectric properties to avoid signal broadening and heat buildup, while operating over a wide temperature range at a consistent frequency. Other essential components of the cellular telephone include ceramic magnetic switches that contain rare earth elements (REs) and indium, and the base stations for the cell phone networks that also use the element indium, as well as tantalum. Each of these minerals has specific properties important to the function of the given component for which substitution of other minerals or their derived materials is presently difficult (see figure on facing page).

The commercial picture for cellular communications is complex. The global market for the ceramic (mineral) materials used in the telephone is relatively small ($400 million to $500 million; Vanderah, 2002); however, these materials are essential for the manufacture of the telephone components, the market for which is approximately 10 times larger. The end-user commercial market is the cellular communication systems (including the base station infrastructure) that make use of the components with a market value on the order of 100 times that of the basic materials. This is the reason that most people are offered a free cell phone if they subscribe to a telephone service.

One of the primary issues in material availability is that of obtaining high-quality titanium dioxide (TiO_2), the basic starting material from which the dielectric heart of the phone (as well as numerous critical components in base stations) is produced. In recent years, suppliers in England, Canada, and Germany have shut down because of environmental concerns associated with chloride processing of the materials (T. Negas, personal communication, March 2007).

Markets for many of the specialized minerals or mineral products in the cell phone are small in that the volume of material needed for cell phones is small. As a result, a new or expanding use significantly increases overall demand for the element and prices can increase significantly. Indium is a prime example. Indium tin oxide, an ingredient used in the production of liquid crystal display products for many applications including cell phones, has come under increasing demand during recent years (short term) and indium prices rose from about $200-300 per kilogram in the late 1990s to more than $800 per kilogram in 2006 (USGS, 2007).

Tantalum and REs are subject to similar increases in demand although their prices have not experienced similar recent increases. Tantalum is essential for the dielectric resonators in 2.2 gigahertz cellular base stations. Although some substitutes for tantalum exist, they result in a loss of performance (i.e., more dropped calls) or shorter battery life. REs are needed for ceramic magnetic switches in the cell phone. Since the only U.S. RE mine operated by Molycorp has been closed, manufacturers of components have become dependent primarily on Chinese suppliers for REs. China is the world's largest exporter of REs.

Cellular telecommunications will not be shut down because of mineral supply restrictions. However, supply restrictions for specific minerals, should they occur, will slow the development of better systems that use restricted minerals. At the same time, higher mineral costs could also motivate the development of new technologies that incorporate mineral substitutes yielding the same performance as the restricted mineral.

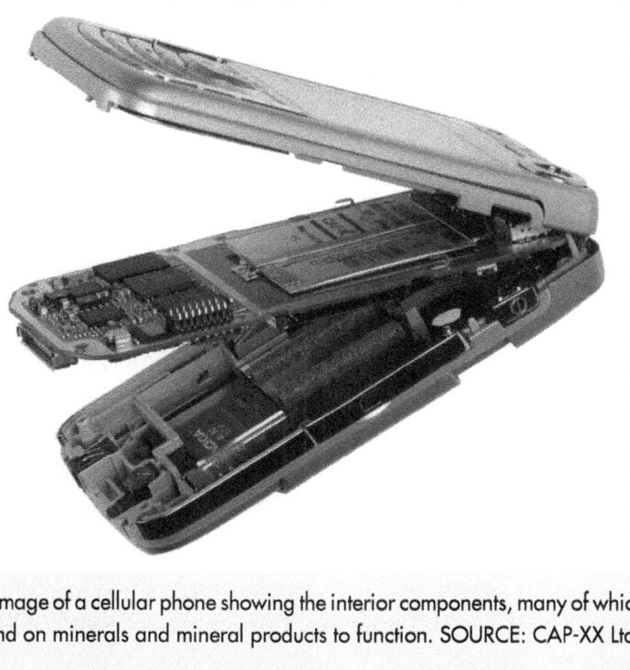

Cutaway image of a cellular phone showing the interior components, many of which contain and depend on minerals and mineral products to function. SOURCE: CAP-XX Ltd.

BOX 1.2
Statement of Task

Understanding the likelihood of disruptive fluctuation in the supply of critical minerals and mineral products for domestic applications, and making decisions about policies to reduce such disruptions, requires thorough understanding of national and international mineral sources, mineral production technology, the key uses of minerals and mineral products in the United States economy, and potential impediments to the mineral supply.

This study will:

1. Identify the critical minerals and mineral products that are essential for industry and emerging technologies in the domestic economy.
2. Assess the trends in sources and production status of these critical minerals and mineral products worldwide.
3. Examine the actual or potential constraints, including but not limited to geologic, technologic, economic, and political issues, on the availability of these minerals and mineral products for domestic applications.
4. Identify the impacts of disruptions in supply of critical minerals and mineral products on the domestic workforce and economy.
5. Describe and evaluate the current mineral and mineral product databases and other sources of mineral information available for decision making on mineral policy issues.
6. Identify types of information and possible research initiatives that will enhance understanding of critical minerals and mineral products in a global context.

reports have included congressionally mandated studies, as well as those conducted at the direct request of federal agencies. Some of the studies have evaluated existing federal programs—for example, those of the U.S. Bureau of Mines (NRC, 1994, 1995a) and the USGS and its Mineral Resources Program (NRC, 1996a, 2001, 2004a)—or have been broader in scope, covering regulatory issues of hardrock mining on federal lands (NRC, 1999), mineral resources and sustainability (NRC, 1996b), mining technologies (NRC, 2002), material flow analysis (NRC, 2004b), the competitiveness of the U.S. mining and metals industries (NRC, 1990), and mineral supply issues for specific end uses such as gas turbine engines (NRC, 1995b). This study differs from those previous reports. The broad study task asks the committee to determine which minerals could be considered critical to the nation and what, if any, additional information and research might be appropriate for the federal government to collect and conduct to mitigate disruptive fluctuations in the supply of critical minerals to key U.S. economic sectors. The audience for the study thus includes not only federal agencies, industry, and research organizations, but necessarily also the general public and decision makers.

The committee notes that several important conclusions and recommendations repeatedly emerged from the previous NRC reports noted above. The committee concurs with these recommendations and underscores the fact that their repetition is an indication that they have yet to be fully implemented. The following list paraphrases some of the overarching conclusions and recommendations from those reports that the committee found most compelling as related to this study:

- The United States is a major user and producer of mineral commodities, and the U.S. economy could not function without minerals and the products made from them.
- The federal government should lead the development of coordinated efforts among academic, private, and federal sectors related to research and information collection on minerals and metals.
- The federal government has a responsibility to conduct and support research and to gather and disseminate information on minerals and metals.
- Market forces alone are not sufficient to meet challenges of sustainability, so the federal government should help facilitate activities that sustain mineral supplies, including exploration, development, technology, recycling, and appropriate environmental protection.
- The federal government should maintain core competence in the knowledge of mineral deposits and related environmental research, as well as information collection, to respond to future national needs.

- Globalization means that mineral resources have become an issue with importance for national security.

The existence, scope, and size of the National Defense Stockpile (NDS) has also often been part of the national discussion of critical or strategic minerals, particularly as the NDS addresses issues regarding which minerals might be considered critical and strategic for the purpose of national security. The present committee has been fortunate in that a complementary study that assessed the need for an NDS was conducted concurrently by the National Materials Advisory Board of the NRC (NRC, 2007). National defense-related minerals issues are thus incorporated in the present report only briefly as one subset of the broader U.S. economic picture; the interested reader is otherwise referred to the NDS report for detailed discussion of defense-related minerals and materials (NRC, 2007).

THE CYCLE OF MINERALS AND MATERIALS

Fossil fuels, with their geologic origins as organic materials, are consumed when burned to generate usable energy. As such, they are destroyed and not available for use later. Such is not the case for nonfuel minerals, which in principle can be recycled after initial use. Thus, minerals and mineral products are available as primary resources (extracted from Earth's crust) and also as secondary resources (recovered from scrap). In addition, for a country or region—as opposed to the planet as a whole—the importation of metals or metal-containing products serves as an additional ("tertiary") resource. Therefore, the availability of minerals, mineral products, and materials in the United States ought to be viewed and evaluated as a cycle of materials similar to some of the groundwork presented in the NRC (2004b) report *Materials Count: The Case for Material Flow Analysis*.

The importance of considering the entire materials cycle in the analysis of nonfuel minerals and mineral criticality can be appreciated by examining the copper cycle shown in Figure 1.1. Of the 2830×10^3 metric tons (Gg; 1 Gg = 10^3 metric tons) of copper entering use, about 70 percent (1987 Gg, or 2235 Gg ore minus 248 Gg ore entering repositories) was primary copper; 16 percent (450 Gg) was secondary copper; and 14 percent (510 Gg imported minus 117 Gg exported) was tertiary copper. About 570 Gg of copper, or about one-fourth as much as was mined domestically in 1994, was either landfilled or dissipated. It is notable that the copper system is not at steady state. While inventories of copper may increase or decrease in a given year at the production stage and the fabrication and manufacturing stage, the total stock of copper in use tends to increase over time, as does the stock of copper in landfills and other repositories. For example, Figure 1.1 shows that in 1994, of the total flow of copper entering use (2830 Gg), 55 percent (or 1570 Gg) represented net additions to the total stock of copper in use in products.

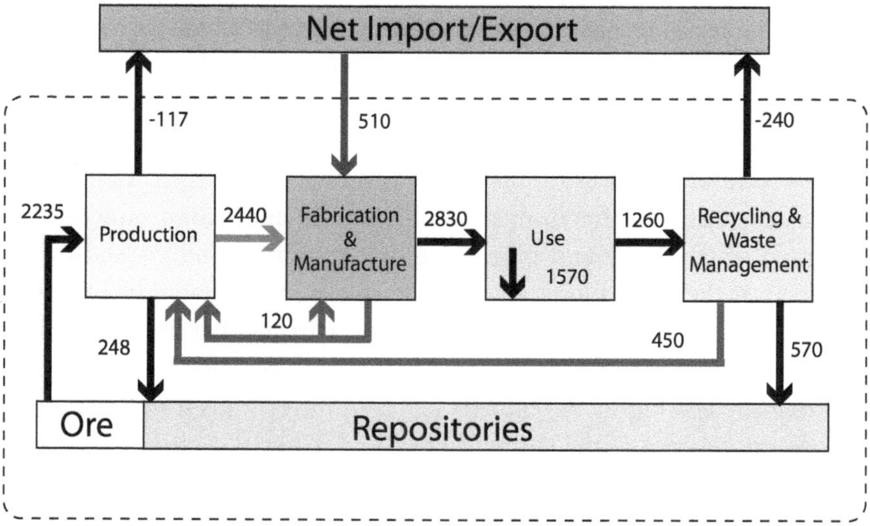

FIGURE 1.1 The U.S. copper cycle in 1994 showing the balance of flows of material for different purposes. The units are thousand metric tons (Gg) of copper. The diagram shows the balance between processing of domestic copper ore (red arrow) plus recycled material (green arrows) and the imported material in semifinished or finished products (blue arrow). Negative numbers and corresponding black arrows represent exports. For many countries, recycled and imported flows are important contributors to the available metal resources. SOURCE: Concept from Graedel et al., 2004.

From this perspective, in-use mineral stock is a future secondary resource and can be regarded with as much interest and analytic rigor as virgin ore bodies. Materials contained in products or applications that reach the ends of their lives in a given time period represent a flow of material that is considered to be available for recycling. On Figure 1.1, for example, about 55 percent of the copper that was removed from service is seen to have been either recycled within the United States (450 Gg) or exported (240 Gg), primarily for scrap processing outside the United States, while 45 percent was discarded to landfills (570 Gg). Material contained in landfills is another secondary resource and could be recovered in the future, but its quality is degraded when mixed with other materials, resulting in a loss of value. The metal content of landfills is difficult to estimate and is highly dependent on the diversion programs in place (e.g., Chandler et al., 1997; Gordon et al., 2006).

Economics and regulations influence the degree to which current needs are met by primary or secondary material (from the global perspective, any tertiary material is fundamentally either primary or secondary). However, a number of other characteristics contribute to the desirability of choosing between different sources for materials, as shown in Table 1.1. In general, primary material benefits from the technological knowledge gained from millennia of discovery and processing, but resource conflicts and other issues can make the exploitation of these stocks problematic. In contrast, secondary or recycled materials possess fewer issues that are potentially problematic, but the collection and reprocessing technologies for those materials are less highly developed. The committee finds it imperative to include the concept of the entire mineral and material cycle in its discussion of mineral criticality.

WHAT IS A CRITICAL MINERAL?

Recognizing that a nonfuel mineral or mineral product can be obtained as either primary or secondary material, what does it mean to say that one of these minerals or mineral products is a *critical mineral*?

In the context of federal communications regarding minerals, the terms

TABLE 1.1 Attributes of Primary and Secondary Materials

	Advantages	Disadvantages
Primary materials	Extensive extraction and processing experience	High energy and water use and air emissions
	Established product specifications and markets	Political disruption a possibility
	Technologies to control impurity levels are well developed	Impacts are "hidden" (often occur in countries other than those in which the material is used)
		Extraction and processing generates high volumes of mine rock, tailings, slags, and residues
Secondary materials	Mostly available in user countries	Collection can be difficult and reprocessing technology is relatively primitive
	Low energy and water use and air emissions	Inappropriate recycling practices pose occupational and environmental health risks in some countries
	Socially and politically acceptable	
	Processing generates low volumes of waste	
	Quality of recycled material is suitable for most applications	Quality of recycled material may not be suitable for some applications

"critical" and "strategic" as mineral or material descriptors have been closely associated, but usually not clearly differentiated. A review of some of these definitions is useful before describing the definition of critical mineral adopted by the committee for this report.

DeYoung et al. (2006) noted that, historically, "strategic materials" in the United States have generally been associated with material availability in times of war or national emergency; the term "critical material" did not enter the federal lexicon until just prior to World War II when it was introduced in the language for the Strategic and Critical Materials Stock Piling Act of 1939 (P.L. 96-41, 1939). The Strategic and Critical Materials Stockpiling Act of 2005 (50 U.S.C. 98 et seq.) defines strategic and critical materials as those that are needed to supply the military, industrial, and essential civilian needs of the United States during a national emergency that are not found or produced in the United States in enough quantities to meet such needs. Specific distinctions between "strategic" and "critical" are not offered in these documents.

The association of the term strategic mineral almost exclusively with national security and military needs or requirements during national emergencies is implicit in the synonyms for "strategic," which include planned, tactical, and calculated. "Critical" in general English usage can refer to something that is vital, important, essential, crucial, or significant. These differences are supported and further refined by definitions in the academic literature suggesting that materials for military uses are strategic, while those for which a threat to supply from abroad could involve harm to the nation's economy are critical (Evans, 1993, in DeYoung et al., 2006). This definition builds on the use of the term critical materials in the context of discussion around the establishment of the National Critical Materials Council in the mid-1980s. Critical materials in this context encompassed any materials—from metals to alloys to composites—on which the economic health and security of the nation resided (Robinson, 1986). A critical material thus has broader connotations than a strategic material, and its definition can be considered to include civilian, industrial, and military applications that could have measured effects on the nation's economy should supply of the material under evaluation become restricted. In accordance with these definitions, a critical material may or may not be strategic, while a strategic mineral will always be critical. This study addresses critical minerals, as opposed to those that may more narrowly be considered strategic, and may differ slightly, for example, from the definition of strategic used by the Industrial College of the Armed Forces (ICAF, 2006).

In the opinion of the committee, a mineral can be regarded as critical only if it performs an essential function for which few or no satisfactory substitutes exist. This dimension of criticality is therefore related to the demand for a mineral that meets very precise specifications required in certain key applications, but it is not simply related to overall demand for all applications. Instead, it reflects economic, social, and other consequences if essential functions cannot be delivered. In addition, a mineral can be regarded as critical only if an assessment also indicates a high probability that its supply may become restricted, leading either to physical unavailability or to significantly higher prices for that mineral in key applications. In turn, the probability of a restriction in the supply of a critical mineral is more likely to be assessed as high if the aggregate demand for key applications represents a relatively large proportion of the overall supply of the mineral that meets the required specifications. Examples presented later in this report also emphasize the distinction between minerals that are *essential* to the economy in certain applications and yet are *not critical*, at least at present, because the risk of supply restriction is low.

In its work, the committee found the concept of a "criticality matrix" to be a useful way to characterize the many variables that influence a mineral's criticality. Determining a mineral's criticality, then, is a means by which decision makers can help alleviate potential impacts of a restriction on the supply of a mineral, or avoid a supply restriction entirely through informed decisions. The matrix concept is developed qualitatively in the next section and in detail in subsequent chapters.

THE CRITICALITY MATRIX

The two important dimensions of criticality are *importance in use* and *availability*. Importance in use embodies the idea that some nonfuel minerals or materials are more important in use than others. Substitution is the key concept here. For example, if substitution of one mineral for another in a product is easy technically, or relatively inexpensive, one can say that its importance is low. In this case, the cost or impact of a restriction in the supply of the mineral would be low. On the other hand, if substitution is technically difficult or is very costly, the importance of the mineral is high, as would be the cost or impact of a restriction in its supply. This concept of importance at a product level significantly includes the net benefits customers receive from using a product—the benefits to human health of nutritional supplements or pollution control equipment, the convenience of cell phones, the durability of an automobile, and so on.

A nonfuel mineral can be important at a scale larger than a product as well as at the product level. A mineral might be important to the commercial success of a company and the company's profitability (importance at a company level). A mineral might be important in military equipment and national defense. Production of a mineral—or products that use the mineral as an input—might be an important source of employment or income for a local community, a state, or the national economy (importance at a community, state, or national level). In all of these cases, the greater the cost or impact of a restriction in supply, which depends importantly on the substitutability of the mineral in question, the more important is the mineral.

Availability is the second dimension of criticality. Fundamentally, society obtains all nonfuel minerals through a process of mining and mineral processing (primary supply). Later, however, in the course of fabrication and manufacturing—and ultimately after products reach the end of their useful lives—society can obtain mineral products through the processing of scrap material (secondary supply). Availability reflects a number of medium- to long-term considerations: geologic (does the mineral exist?), technical (do we know how to extract and process it?), environmental and social (can we extract and process it with a level of environmental damage that society considers acceptable and with effects on local communities and regions that society considers appropriate?), political (how do policies affect its availability both positively and negatively?), and economic (can we produce a mineral or mineral product at costs consumers are willing and able to pay?). In addition, it is important to consider the reliability or risk of supply in the short term. Is the nation vulnerable to unexpected disruptions in availability due to, for example, import dependence, market power in the hands of a small number of powerful producers, thin or small markets that are unable to respond quickly to changing circumstances, or significant changes in public policy that cut off supply or increase costs?

In both dimensions of criticality, time is an important consideration. In the short term (period of a few years or less) or the medium term (less than 10 years), both mineral users and producers generally are less able to respond quickly or effectively to changing market conditions than over longer time periods. Even within a particular period, however, some minerals will be more important in use and more vulnerable to supply disruptions than other minerals. For a given adjustment period (short term to long term), the critical minerals are those that are relatively difficult to substitute and are subject to supply risks.

Figure 1.2 illustrates this concept of criticality and the criticality matrix. The vertical axis embodies the idea of importance in use and represents the impact of supply restriction. The horizontal axis embodies the concept of availability and represents supply risk. One can evaluate a mineral's criticality by evaluating its importance in use and its availability, and locating it on the figure. The degree of criticality increases as we move away from the figure's origin, as shown by the arrow and the increased shading. Mineral A, for example, is more critical than mineral B. In this sense, criticality is appropriately considered a "more-or-less" issue rather than an "either-or" issue. That is, minerals exhibit differing degrees of criticality depending on the circumstances. Some minerals are more critical than others; it is a matter of degree rather than absolutes. To be sure, some mineral users or government officials may want to create a list of critical minerals, implying that minerals not on the list are not critical, for purposes of planning or policy making. The committee has not created a definitive list of critical minerals because it did not have the time or resources to assess all possible critical minerals. Rather, Chapter 4 illustrates how the matrix can be used and suggests several candidate minerals for criticality. The committee used a combination of quantitative measures and expert (qualitative) judgment in implementing the matrix methodology.

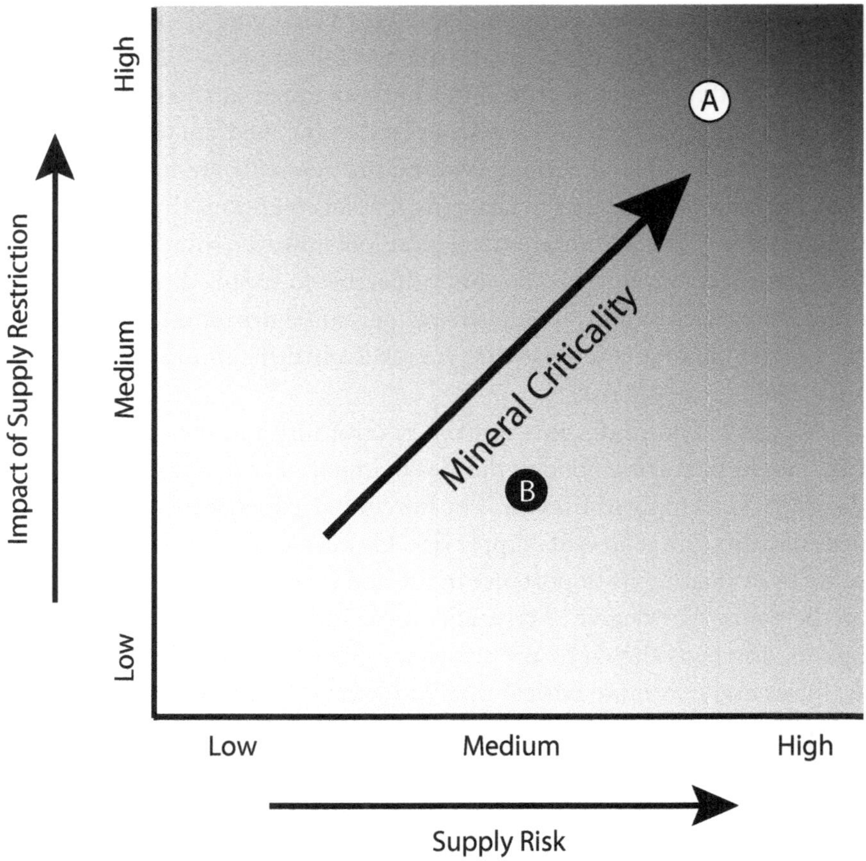

FIGURE 1.2 The criticality matrix, as established in this report, allows evaluation of the criticality of a given mineral. A specific mineral or mineral product can be placed on this figure after assessing the impact of restriction on the mineral's supply should it occur (vertical axis) and the *likelihood* of a supply restriction (horizontal axis). The degree of criticality increases as one moves from the lower-left to the upper-right corner of the figure. In this example, mineral A is more critical than mineral B. More specific descriptions of the parameters used to evaluate mineral supply restrictions and their impacts are presented in Chapters 2-4.

COMMITTEE PROCESS

To address the statement of task and establish report recommendations, the committee reviewed relevant NRC reports; information submitted by and requested from external sources, including two open meetings (see Appendix B); other published reports and literature; and importantly, information from the committee's own experience. The committee held three meetings in Washington, D.C., two of which were at the National Academies' Keck Center and one at the National Academy of Sciences Building (Appendix B). The first meeting in December 2006 included a dialogue with the study's sponsors and other federal agency participants. This meeting also allowed the scope of the concurrent NRC study on the NDS to be discussed, and the two NRC studies have functioned in a complementary and participatory manner during the open sessions of one another's meetings (see also NRC, 2007).

The second meeting of the committee, in March 2007, was the main information-gathering session of the study and consisted of a 2-day open session with four panel discussions. The meeting gathered a spectrum of panelists representing mineral product "users" from the private sector, and individuals who could speak about the sources for minerals and mineral products, the various potential constraints on mineral supply, and the data, research, and information on mineral availability and prices that can be provided to the public. One of the main questions the committee posed to the panelists was with regard to definitions of the term *critical mineral* as might apply to specific manufacturing sectors or consumers and users of mineral-containing products. These definitions weighed into the committee's deliberations to frame the concept of mineral criticality in this report. The final meeting of the committee was a closed session held in May 2007 at which time the recommendations were reviewed. Throughout the study process, the committee also received valuable input through informal interviews with various professionals associated with mineral information, use, and availability and was greatly supported in its work by voluntary contributions from a spectrum of interested individuals from across the country.

REPORT STRUCTURE AND CONCLUDING REMARKS

Having set the stage with this chapter, the committee has organized its analysis in the remainder of the report as follows: Chapter 2 examines the vertical axis of the criticality matrix—the importance of minerals in use—through examples that demonstrate specific applications of minerals and materials in some important U.S. industry sectors and the degree of mineral substitutability in these applications. Chapter 3 examines the horizontal axis of the matrix—the availability and reliability of mineral supply—to describe more completely the numerous factors that can affect mineral supply from short- through long-term periods. Chapter 4 demonstrates application of the matrix to evaluate mineral criticality by examining 11 mineral candidates. The minerals and their applications cross many more industry sectors than the four examined in detail in Chapter 2 and serve to underscore the ubiquitous applications for minerals in everyday life. Chapter 5 presents an overview of the federal data gathering, information, and research efforts appropriate for making informed decisions about minerals, in general, and critical minerals, in particular. Finally, Chapter 6 presents the report's main conclusions and recommendations.

The committee notes that the statement of task refers to "minerals and mineral products"; to streamline the text, the remainder of this document uses the term "minerals" to encompass nonfuel minerals and mineral products and materials, as well as metals. Key terms are otherwise defined in the text and collected together in the glossary (Appendix C). The committee is not consistent in its use of metric or imperial units; instead the report uses the units most commonly applied by the relevant organization or industry in each case and provides conversions where appropriate. While the report focuses on nonfuel minerals, the committee does, in a very limited manner, consider uranium production and use because uranium often occurs in association with other metallic minerals and demonstrates the importance of a range of minerals and mineral-based products without which it would not be possible to maintain or increase nuclear power generation.

Informed planning to maintain and enhance domestic economic growth requires knowledge of potential resource disruptions. Many existing and emerging technologies require nonfuel minerals that are not available in the United States. Thus, although market forces influence some aspects of the balance between supply of and demand for nonfuel minerals and mineral products, various factors including global mineral distribution; mineral discovery, extraction, and processing; and new technologies and applications for minerals also influence the supply of these minerals to manufacturers and their incorporation in consumer goods. This report is designed to give federal agencies, policy makers, industry, academia, and the general public a framework in which to evaluate critical minerals and to indicate which types of data and research are appropriate to help ensure continuing mineral supplies and to develop suitable substitutes.

REFERENCES

Chandler, A.J., T.T. Eighmy, J. Hartlén, O. Hjelmar, D.S. Kosson, S.E. Sawell, H.A. van der Sloot, and J. Vehlow, 1997. *Municipal Solid Waste Combustor Residues: The International Ash Working Group, Studies in Environmental Science.* Amsterdam: Elsevier Science, 67.

DeYoung, J.H., Jr., L. McCartan, and J. Gambogi, 2006. What's been (and what will be) strategic—My metal or your paint? In Reid, J.C. (ed.), *Proceedings of the 42nd Forum on the Geology of Industrial Minerals: Information.* Circular 34. Raleigh: North Carolina Geological Survey.

Evans, A.M., 1993. *Ore Geology and Industrial Minerals—An Introduction* (3d ed.). Oxford, U.K.: Blackwell Science, p. 10.

Gordon, R.B., M. Bertram, and T.E. Graedel, 2006. Metal stocks and sustainability. *Proceedings of the National Academy of Sciences* 103(5):1209-1214; originally published online Jan 23, 2006.

Graedel, T.E., D. van Beers, M. Bertram, K. Fuse, R.B. Gordon, A. Gritsinin, A. Kapur, R. Klee, R. Lifset, L. Memon, H. Rechberger, S. Spatari, and D. Vexler, 2004. The multilevel cycle of anthropogenic copper. *Environmental Science & Technology* 38(4):1242-1252.

ICAF (Industrial College of the Armed Forces), 2006. AY2005-2006 Industry Study, Final Report, Strategic Minerals. National Defense University, Washington, D.C., 26 pp.

NRC (National Research Council), 1990. *Competitiveness of the U.S. Minerals and Metals Industry.* Washington D.C.: National Academy Press, 160 pp.

NRC, 1994. *Research Programs for the U.S. Bureau of Mines: First Assessment.* Washington, D.C.: National Academy Press, 130 pp.

NRC, 1995a. *Research Programs of the U.S. Bureau of Mines 1995 Assessment.* Washington, D.C.: National Academy Press, 138 pp.

NRC, 1995b. *High-Purity Chromium Metal: Supply Issues for Gas-Turbine Superalloys.* Washington, D.C.: National Academy Press, 60 pp.

NRC, 1996a. *Mineral Resources and Society: A Review of the U.S. Geological Survey's Mineral Resource Surveys Program Plan.* Washington, D.C.: National Academy Press, 99 pp.

NRC, 1996b. *Mineral Resources and Sustainability Challenges for Earth Scientists.* Washington, D.C.: National Academy Press, 26 pp.

NRC, 1999. *Hardrock Mining on Federal Lands.* Washington, D.C.: National Academy Press, 260 pp.

NRC, 2001. *Future Roles and Opportunities at the U.S. Geological Survey.* Washington, D.C.: National Academy Press, 192 pp.

NRC, 2002. *Evolutionary and Revolutionary Technologies for Mining.* Washington, D.C.: National Academy Press, 85 pp.

NRC, 2004a. *Future Challenges for the U.S. Geological Survey's Mineral Resources Program.* Washington D.C.: The National Academies Press, 153 pp.

NRC, 2004b. *Materials Count: The Case for Material Flow Analysis.* Washington D.C.: The National Academies Press, 144 pp.

NRC, 2007. *Managing Materials for a 21st Century Military.* Washington D.C.: The National Academies Press, 144 pp.

Robinson, A.L., 1986. Congress critical of foot-dragging on critical materials. *Science* 234(4772):20-21.

USGS (U.S. Geological Survey), 2007. *Mineral Commodity Summaries 2007.* Reston, Va.: United States Geological Survey, 195 pp.

Vanderah, T.A., 2002. Talking ceramics. *Science* 298(5596):1182-1184.

CHAPTER 2

Minerals and Materials Uses in the United States

NONFUEL MINERALS AND SOCIETY

The nation recognizes the importance of an adequate supply of fossil fuels—oil, natural gas, and coal—which have formed the energy base for much of Europe and North America for the past two centuries. As we have experienced recently, changes in availability, or the perceived availability, of these fossil fuel minerals generate significant price fluctuations and thereby impact the full range of individual and industrial energy consumers in the United States where current technologies do not allow massive substitution of renewable or alternative energy sources for fossil energy. Less visible is our reliance on the large number of minerals that are fundamental ingredients in the manufactured products used in the United States—from cell phones and flat-screen monitors to paint and toothpaste. A large number of the products manufactured or used domestically contain natural mineral resources that either are mined in the United States or are imported from abroad as raw materials or as fully manufactured or semifinished products. Estimates suggest that current lifestyles in the United States require per capita annual consumption of more than 25,000 pounds (11.3 metric tons) of new nonfuel minerals to make the things that we use every day (MII, 2007a).

In addition to their intrinsic, practical value as part of a manufactured product, minerals also have a significant general value to the U.S. economy from both a financial standpoint and an employment standpoint (discussed in some detail at the close of this chapter). Globally the demand for minerals is also important and is increasing. In the emerging economies of countries such as China and India, where industrial output has surpassed

BOX 2.1
Emerging Economies: An Example in China

The emergence and growth of several foreign economies have been of interest to U.S. industries and economic analysts for a number of years. Recent international media attention to the very rapid rates of economic and industrial growth particularly in China and India, relative to Europe, North America, and Japan, indicates that the emerging economies and their global influence are also gaining public interest. Nonfuel minerals factor directly into this situation as key inputs to continued industrialization and manufacturing output for emerging economies, as these nations satisfy both their own, growing domestic consumer needs and the large international demand for their exported products.

As one example, China has become a leading global consumer of products such as cell phones, televisions, and refrigerators and will soon surpass the United States in the consumption of personal computers and automobiles in absolute terms (McCartan et al., 2006). The U.S. trade deficit with China indicates that China is not only a consumer nation, but also a producer of many types of goods for export. The increase in consumer purchases in China has been fueled by industrialization that has triggered a better national standard of living, which in turn feeds further demand for manufactured goods and the infrastructure to power and use those goods. The same can also be said of emerging economies in other nations and represents the path of industrialization that was followed during the past century by North America, Japan, and Europe, among others.

China is endowed with abundant natural mineral resources. Whereas in 1973 the United States was the world's leading nonfuel mineral producer, today China is the world's leading producer and consumer of minerals, but it no longer has the domestic capacity to satisfy its demands for minerals such as iron ore, nickel, copper, and cobalt (McCartan et al., 2006; Reynolds, 2007). China's dominance as a global mineral supplier, coupled with its own demand for minerals, gives it a very direct global market influence on the supply of minerals. In 2004, China also increased its own mineral production capacity and apparthat of Europe or the United States in many sectors, increased standards of living have encouraged greater demands for consumer goods containing processed minerals (Box 2.1). Furthermore, the ability to design and manufacture new materials and to characterize, predict, and exploit the chemical and physical properties of minerals as components of those materials

ently issued 13,000 exploration licenses and 40,000 mining permits. The same year, more than half of China's direct investments in foreign nations went to mining projects (McCartan et al., 2006).

China's demand for minerals affects both the "traditional" and the "emerging" materials markets. With more traditional materials such as steel, Chinese steelmakers compete with Japanese and American steelmakers for raw iron ore (primary) and steel scrap (secondary) supplies globally. Aside from direct influences on the purchase and use of feedstock for U.S., Japanese, and Chinese steelmakers, the effects of greater competition for iron ore and steel scrap will also impact the industries supplied by steelmakers and may affect issues ranging from construction costs for new power plants (Wald, 2007) to the selection or development of new automobile designs. China's growing mineral demand also affects emerging materials and technologies, such as those requiring rare earth elements (REs). China is the world's leading producer and consumer of REs, holding a 97 percent market share of processed REs (USGS, 2007a). The uses of REs are discussed in more detail later in this chapter and in Chapter 4.

As mineral demand has increased in China and other emerging economies in this century, mineral prices have risen—the third significant boom in mineral commodity prices since World War II. Emerging economies and their needs for minerals underscore the fact that the U.S. economy is affected by global production and consumption demands, the satisfaction of which requires manufacturing, industrial capacity, and supporting infrastructure. Nonfuel minerals are key inputs to the manufacturing process both domestically and abroad, and their availability to U.S. industry and manufacturers has emerged as a much more competitive process than previously. Consistent, accurate, and current information on global mineral trends is a foundation for U.S. domestic industry and government to make effective plans and decisions regarding mineral supply.

SOURCES: McCartan et al., 2006; Reynolds, 2007; Wald, 2007; USGS, 2002, 2007a.

continues to advance and improves the technical performance, durability, and reliability of products. These improvements deliver greater value to businesses and consumers, and in some cases to the environment, but also serve to increase demand for specific types of minerals.

Understanding the importance of minerals in the products from different sectors of the U.S. economy forms the basis for the vertical axis of the criticality matrix described in Chapter 1 (Figure 1.2). As discussed in Chapter 1, end uses or applications for minerals will have varying levels of importance depending on the demand for that particular end use. *Importance in use* carries with it the concept that some minerals will be more key or fundamental for specific uses than others, depending on a mineral's chemical and physical properties. Focusing on a mineral's properties emphasizes the role of *substitutability* as a factor in determining criticality. A mineral that has ready substitutes—that is, substitute minerals that provide similar properties or performance at a comparable price—is less critical than a mineral with few substitutes. The concept of substitutability has both technical and economic meaning. A technical substitute provides the same or similar performance compared to the mineral it replaces. To be an economic substitute, however, the mineral needs to provide the same or similar performance at similar or lower costs, or better performance at the same costs. Ease of substitution, in turn, determines the degree of a mineral's importance in use—the vertical axis of the criticality matrix (see also Figure 1.2). A useful concept to consider also is that the demand for minerals is a derived demand. That is, users demand minerals for the chemical and physical properties they and the elements extracted from them provide. Copper, for example, provides electrical conductivity. Zinc, when used for galvanized steel, provides corrosion protection. Aluminum provides light weight, coupled with strength and metallurgical formability.

This chapter begins with an introduction to minerals and the properties that make them important for specific applications. The chapter then describes key uses of the minerals in four major industrial sectors—automotive, aerospace, electronics, and energy—before concluding with a general discussion on the potential impacts of restrictions in the supply of minerals on the domestic economy. These four sectors were chosen as useful examples for purposes of the discussion on mineral criticality, but similar arguments could be made for other important sectors such as health care, construction, utilities, or the transportation infrastructure. All sectors of the economy rely on the services provided by minerals and the committee was limited in its time and resources to selecting a few industry examples. Discussion of the critical mineral candidates in Chapter 4 includes examples of mineral applications from many additional industry sectors.

CHEMICAL AND PHYSICAL PROPERTIES OF MINERALS

Chemical elements are the building blocks of all minerals. Of the 111 authenticated elements in the periodic table, the first 92 (up to uranium) occur naturally (Appendix D). Of these, the ones relevant to this study are those that occur as solid minerals. Few are found in nature in their pure form; "native copper" and the precious metals gold, silver, and platinum, for example, are exceptions in that they occur in pure metallic form (although not as pure elements). Some elements occur as simple compounds with oxygen, for example, iron oxides (a key component of steel), titanium dioxide (a key component in pigments), or silicon dioxide (quartz, or silica, which is important in many computing applications), or they may combine with several other elements to form one of many naturally occurring minerals. The wide range of elements (or minerals or compounds in which they are found) can be classified broadly into one of several basic categories that largely represent the key mechanical, electrical, magnetic, or optical properties specific to similar elements in that group (Table 2.1).

The unique properties of elements and the minerals in which they are found allow some minerals to be used directly on extraction from the Earth, or after relatively limited processing, whereas others must undergo extensive transformation and processing to produce metals, metal alloys, inorganic metal compounds, electronic materials, manufactured parts, and construction materials. Metals are used almost exclusively in the form of alloys, in which one metallic element is combined with one or more metallic or nonmetallic elements to improve properties such as strength,

TABLE 2.1 Classification and Properties of Some Elements

Group	Elements
Precious metals including platinum group metals (PGMs)	Gold, silver; PGMs: ruthenium, rhodium, platinum, palladium, osmium, iridium
Base metals and ferrous metals	Copper, nickel, aluminum, tin, zinc; iron (ferrous metal)
Rare earths (lanthanide series only[a])	Lanthanum, cerium, praseodymium, neodymium, promethium, samarium, europium, gadolinium, terbium, dysprosium, holmium, erbium, thulium, ytterbium, lutetium
Transition (T), alkaline and alkaline earth (A), and other metals (O), metalloids (Md) and nonmetals (NM)	T[b]: Scandium, titanium, vanadium, chromium, manganese, cobalt, zirconium, cadmium, niobium, molybdenum, tantalum, tungsten, rhenium, mercury A: Lithium, sodium, potassium, beryllium, magnesium, calcium, strontium, barium O[c]: Gallium, indium, lead Md: Boron, silicon, germanium, arsenic, antimony, tellurium NM: Carbon, oxygen, phosphorus, sulfur, selenium

NOTE: Properties of the minerals have been highly simplified and serve as basic examples of the concept of importance in mineral chemical and physical properties for applications.

light weight, conductivity, durability, or corrosion resistance. Minerals that can be extracted and used after relatively limited processing include sand, gravel, dimension stone, agricultural limestone, and clays (Box 2.2). Another category of materials includes ceramics, which can be broadly defined as inorganic nonmetals and includes such diverse materials as silicon dioxide-based glasses, crystalline oxides (e.g., aluminum oxide, Al_2O_3),

Properties
Scarce, corrosion and tarnish resistant, malleable, ductile, high luster and high electrical conductivity
Reasonably abundant, high electrical conductivity, high heat conductivity, some may corrode fairly easily
Extremely diverse metallurgical, chemical, catalytic, electrical, magnetic, and optical properties (see also Chapter 4)
T: Ductile and malleable; conduct electricity and heat; some are magnetic A: Malleable, ductile, and are good conductors of heat and electricity; softer than most other metals; very reactive O: Ductile and malleable; high density Md: Properties of both metals and nonmetals; some such as silicon and germanium are semiconductors making them useful in computers and calculators NM: Often exist as gases; not able to conduct electricity or heat very well; very brittle

[a]REs also include the actinide series which is not discussed further here. Scandium and yttrium are sometimes also considered REs.

[b]Some transition metals are base metals (nickel, copper, zinc, iron) or precious metals (silver, gold, palladium, platinum, osmium).

[c]Some nontransition, nonalkaline metals are also base metals (aluminum, tin).

carbides (e.g., silicon carbide, SiC), and nitrides (e.g., aluminum nitride, AlN); carbon-based products such as graphite; and some semiconductor materials such as zinc selenide (ZnSe) and gallium arsenide (GaAs). Ceramic applications can range from traditional products such as whitewares and structural clays to recent product developments in cellular communication, aerospace, and health care.

BOX 2.2
Aggregates

In 2006, the United States used 2.69 billion metric tons of crushed stone, sand, and gravel (referred to here as aggregates; USGS, 2007b). Aggregates used for construction perform three major functions. The first is as a structural component of portland cement concrete (often called ready-mix concrete) and bituminous cement concrete (often called asphalt), used in the construction of streets, highways, and parking lots. Ready-mix concrete is also used in the construction of residential and nonresidential buildings, bridges, dams, and water and sewage treatment plants. The second general function is applications in which their high bulk specific gravity is important for uses such as erosion control and nuclear reactor containment structures. Aggregates perform a third general function in which they are used without addition of other materials to take advantage of their properties such as strength (for road base, under concrete slabs, foundations), permeability (used behind retaining walls, French drains, sewage treatment plants, septic drain fields), and volume (backfill in pipe trenches).

Aggregates are ubiquitous in the United States and their applications are a fundamental part of the nation's infrastructure. While these minerals are very important and may be critical to local industries and governments, their domestic supply is relatively secure and they would fall in the criticality matrix of Chapter 1 as low-risk, high-impact minerals. On the supply side, this is illustrated by the fact that the 3 billion tons of aggregates produced in the United States in 2006 derived from more than 10,000 locations operated by more than 6000 companies. On the use side, although aggregates have limited or no readily available or suitable substitutes, the impact of a supply disruption at any one or even several operations is likely to be minimal because of the proximity or number of other potential supply points in the country.

However, two situations—one in Florida and one in California—illustrate the idea of aggregate criticality within a specific geographical area, which could in turn affect that local (or state) economy. In March 2006 in Miami-Dade County, Florida, a district court judge ruled that permits to allow the mining of limestone and lime rock had been issued improperly in 2001 by the U.S. Army Corps of Engineers. The ruling stated that the Corps would have to conduct further studies to substantiate its decision that mining in the affected acres would not damage wetlands, degrade groundwater, or threaten endangered species. The affected operations in the area, called the Lake Belt, represent approximately 56 percent of the total aggregate production in Florida. If an injunction is issued to prohibit further mining until the completion of a new environmental impact study, the price of aggregate could rise an estimated 61 percent to 124 percent (Morrell, 2006). This estimate assumes that replacement aggregate can be obtained from existing sources, some of which are 500 to more than 1000 miles away. The high bulk density of aggregates makes them very expensive to transport. In addition, the existing transportation infrastructure would have to adjust quickly to handle the increased shipments of aggregates from distant, as opposed to local, sources. To date, a Florida judge has ruled to shut down 3 of 12 permits for operations in the area (Associated Press, 2007), one of which is the largest in the United States. The Lake Belt example illustrates the potential short-term consequences of a restriction in aggregate supply.

The second situation in California relates to the long-term reliability of supply and results from years of high production and consumption coupled with strong environmental resistance to permitting of new aggregate operations. A report (Kohler, 2002) compared the projection of permitted reserves on January 1, 2005, to actual permitted reserves as of January 1, 2001, and indicated a decline in permitted reserves from 6.8 billion to 4.3 billion tons over that period. Projected demand for the same period went from 12.0 billion to 13.5 billion tons. While the permitted reserves will last an average of 17 years, 4 of the 31 aggregate study areas—North San Francisco Bay, Sacramento County, Fresno County, and northern Tulare County—are projected to have less than 10 years of permitted aggregate resources remaining. The California Geological Survey has expressed concern that California will increasingly experience shortages in the future. Alternate supply sources exist, but with California aggregate consumption reaching 235 million tons in 2005, imports from Canada and Mexico would have to be considered to overcome some of the deficit in production from permitted reserves.

In each of the above cases, aggregate supply to fill the anticipated demand in the event of restrictions in local supply will have to come from neighboring states or be imported from Canada, Mexico, or Caribbean islands. With a commodity such as aggregate that has a very high cost of transportation and a practical limitation to distributing large quantities of material from ports to the point of consumption, imports are difficult to envision as realistic alternative sources of aggregate supply. The basic issue to consider is that local regions or states can have their aggregate supplies restricted to the point at which even a relatively "common" material such as aggregate can become critical. With aggregates, situations can develop that cause a supply interruption that cannot easily or quickly be remedied. While each state is monitoring the status of its aggregate supply and demand, and their balance relative to environmental concerns, the states and aggregate operating companies require continuous, accurate, and unbiased information on aggregate production and supply throughout the country and internationally to be able to make appropriate decisions to meet a state's aggregate demands.

MINERAL USES

Most minerals are used in long-lived products that provide benefits to consumers over extended periods. The most important sectors of the economy for mineral demand are transportation, including automobiles and airplanes; capital equipment such as industrial machinery; residential and commercial construction; and consumer durables such as washing machines, refrigerators, cellular telephones, and televisions. This section presents some of these common and widely used products and their constituent minerals from four key sectors of the U.S. economy. The discussion emphasizes chemical and physical properties of minerals and their substitutability and the dynamic or changing demand for minerals through time. Substitutability and dyna-

BOX 2.3
Defense

The defense of the United States presents its own special issues in terms of the availability of minerals and materials. This particular set of minerals is viewed by the committee as "strategic" in nature (see Chapter 1 for definitions). Some minerals that are particularly strategic to the defense sector include some of the rare earths (REs) as well as rhenium, cobalt, and beryllium.

The REs have various applications, including neodymium in high-strength magnets and as dopant for lasers, samarium in samarium-cobalt magnets, yttrium in laser rods and superalloys, and scandium in aluminum alloys and refractory ceramics. Rhenium is a rare metallic element of importance to the defense community because of its contribution to the properties of high-temperature alloys and coatings (Sloter, 2007). Rhenium is obtained almost exclusively as a by-product of the processing of a special type of copper deposit known as a *porphyry copper* deposit. Specifically, it is obtained from the processing of the mineral molybdenite (a molybdenum ore) that is found in these copper deposits. Even though the United States has significant rhenium reserves, the majority of this mineral consumed in the United States has been imported, primarily from Chile and Kazakhstan (USGS, 2007a).

Beryllium is perhaps the best example of a strategic defense mineral because of its unique combination of mechanical and nuclear properties and issues with its production and mism are important factors to consider when evaluating the importance of end use in determining mineral criticality. Defense applications constitute a special sector in terms of national security because they require products from many industrial sectors. The committee addresses some of the mineral needs of the defense sector briefly in Box 2.3 and refers the reader to the more detailed treatment of the defense sector and strategic minerals in the report of the National Research Council (NRC) Committee on Assessing the Need for a Defense Stockpile (NRC, 2007). The vast numbers of uses for the minerals in the United States are too numerous for the committee to describe in detail, and only a few examples are provided in this report of the common products in which minerals are components.

supply. It is a lightweight metal possessing relatively high stiffness and strength, making it a material of choice for a wide range of defense systems including sensors, aircraft, missiles, satellites, and nuclear warheads.

The U.S. Defense Department is concerned because Brush Wellman, Inc. (BWI), the only U.S. producer of beryllium, closed its production facility in the year 2000 because of economic and health and safety issues. Since then, BWI has relied on a dwindling supply of beryllium ingot that it purchased from the National Defense Stockpile. The stockpile also has uncommitted beryllium inventories of hot-pressed powder billets being held in reserve, but these could extend the depletion date for only a few years (*http://www.acq.osd.mil/ott/dpatitle3/projects/bp.htm*).

Other supplies of beryllium are obtained from Kazakhstan, but the purity of these imports is insufficient for a number of critical defense applications. A report to Congress noted that it would take a minimum of 3 to 5 years to complete a new primary beryllium facility. The report also indicated that even if the purities of the imports from Kazakhstan were to reach acceptable levels, the risks of sole source dependence on that country for production would be unacceptable. Authorization for BWI to transfer its beryllium manufacturing technology to Kazakhstan to improve the quality of its imports introduces other risks—namely, the sale of pure beryllium metal to third world countries seeking to produce nuclear weapons (*http://www.acq.osd.mil/ip/docs/annual_ind_cap_rpt_to_congress-2005.pdf*).

Automotive

The automotive industry has been an important sector of the U.S. economy, especially in regions where this industry is concentrated. Moreover, from the perspective of households, automobiles are integral to our way of life. The automotive industry uses large amounts of material each year and a modern automobile can contain at least 39 different minerals. Table 2.2 lists some of the primary minerals and metals, their content by weight in an average automobile, and the properties that make the minerals useful for specific parts or functions. Each of these minerals performs a special function alone or in alloys, to enhance performance, durability, safety, and comfort, or to reduce environmental impact or weight, all of which are important features to the automobile owner and, thus, to the industry.

In contrast to the list of minerals in today's automobiles (Table 2.2), the earliest automobiles contained a very small suite of materials, including steel, wood, rubber, glass, and brass. The properties of steel made it a natural replacement for wood in cars through time. Vanadium-steel alloy used by Henry Ford in the first Model T autos rendered the material lighter and stronger than the standard steel of the early 1900s and was an early example of material modification on a mass production scale to improve the overall technical performance and durability of the automobile (Gross, 1996).

The desire for increased strength with decreased weight encouraged the development and use of various steel alloys containing a variety of important minerals including molybdenum, chromium, nickel, and manganese, in addition to vanadium. Molybdenum is important as an alloying element in stainless and other steels and imparts strength and toughness to the material. It is worth noting that a sharp increase in the price of molybdenum during the late 1970s led to the development of high-strength, low-alloy (HSLA) steels, which required less or no molybdenum. Chromium alloyed with steel makes it both corrosion resistant and harder. Nickel imparts strength at elevated temperatures, and manganese is also important as a hardener of steel or when alloyed with aluminum and copper. Vanadium hardens and strengthens iron when alloyed with it—of particular importance now in the manufacture of piston rods and crankshafts.

TABLE 2.2 Some Minerals and Their Weights and Properties in Today's Automobile

Mineral	2006 Weight (pounds/ kilograms)[a]	Property
Iron and steel	2124/963	High strength, durability (frame, motor)
Aluminum	240/109	Light weight (frame, motor)
Carbon	50/23	Bond strengthener (tires and other rubber parts)
Copper	42/19	Electrical conductivity
Silicon	41/19	Bonding properties (windshields and windows)
Lead	24/11	Conductor (storage batteries)
Zinc	22/10	Galvanizer; strengthens in metal alloys (die cast parts and galvanized metal)
Manganese	17/8	Hardens as metal alloy
Chromium	15/7	Corrosion resistance and hardness as metal alloy
Nickel	9/4	Strength at elevated temperature and corrosion resistance as metal alloy
Magnesium	4.5/2	Alloying element with other metals such as aluminum
Sulfur	2/0.9	Strengthens rubber tires
Molybdenum	1/4.5	Strength and toughness as metal alloy
Vanadium	<1/<0.45	Strengthens, hardens, lighter weight as metal alloy
Platinum	0.05-0.10 troy ounce/ 1.5-3.0 grams	Catalytic properties (catalytic converters)

NOTE: In addition to the minerals and metals listed above, the average automobile also contains trace amounts of phosphorus, niobium, antimony, barium, cadmium, cobalt, fluorspar, gallium, gold, graphite, halite, limestone, mica, palladium, potash, strontium, tin, titanium, and tungsten. The category "iron and steel" includes cast iron (435 pounds), conventional steel (1382 pounds), high-strength, low-alloy steel (263 pounds), and stainless steel (45 pounds); rubber (140 pounds) and plastics (250 pounds) also constitute an enormous proportion of the total composition of an automobile.

[a]MII (2007b).

Aluminum and steel overlap in such applications as the frame or engine. The average weight of an automobile is 2600 to 3000 pounds. The desire to reduce weight and contribute to improved fuel economy has led to an increased use of aluminum, which is less dense than steel. The amounts of aluminum and steel, particularly as part of the frame of an automobile, are thus evaluated not as a strict function of direct substitutability, but also as a function of the desired features of automobile design.

Like aluminum, the demand for copper in automobiles has increased through time in response to its properties as an excellent electrical conductor. In 1948, the average family car contained only 55 copper wires with a total combined length that averaged 45 meters (150 feet). Continuing improvements in electronics and the consumer's desire for power accessories in automobiles led to today's automobiles, which contain up to 1500 copper wires that total about 1.6 kilometers (1 mile) in length. A typical midsized automobile contains about 22.5 kilograms (50 pounds) of copper, including some 18 kilograms (40 pounds) of electrical components (*http://www.copper.org/education/c-facts/c-trans_industry.html*). Because of a greater amount of electrical wiring, fuel-efficient hybrid cars require more copper than conventional cars—potentially up to an added 12 kilograms (circa 26 pounds; Stablum, 2007), as well as larger quantities of other metals such as cobalt, nickel, and/or lithium with electrochemical and thermal properties that make them important as components (electrodes) of the rechargeable battery system for hybrid vehicles (either lithium ion or nickel-metal hydride batteries) (Chavasse, 2005).

The platinum group metals (PGMs) are used primarily for their excellent catalytic properties; their resistance to chemical attack, wear, and tarnish; stable electrical properties; and stable behavior at temperature. Of the PGMs, platinum, palladium, and rhodium are crucial to the operation of automotive catalytic converters and, at a time when reduction in air pollution is a primary goal, are therefore crucial to the manufacture of the automobile itself (Herring, 2007). Platinum in the catalytic converter leads to a reduction in carbon monoxide and hydrocarbon emissions. Palladium is an adequate substitute for platinum in gasoline catalytic converters, but in diesel engines only platinum will work as a suitable catalyst. Rhodium, for which there is no known substitute in catalytic operation, is used to reduce NO_x emissions.

In assessing the impact of these materials on the automotive industry one must consider not just availability, but also price. The PGMs are among the rarest metals in the world. The price of platinum has increased about sixfold in the past 3 years, creating additional considerations for the automotive industry as well as other applications, including catalysts, films for electronic circuits, jewelry, and dental crowns.

The rare earth elements (REs) are important to a number of applications. Catalytic converters will not operate without cerium and lanthanum, for which no substitutes are known. One of the REs, neodymium, also plays a significant role as a component in high-strength magnets, important to the automotive industry as a major part of power windows. RE use in catalytic converters ranks highest in proportion to the total annual market for REs in the United States, garnering 46 percent of the total RE market annually (USGS, 2007a) (see also Chapter 4).

Aerospace

The aerospace industry is another sector that depends on the availability of critical materials. Aerospace manufacturing is important to the economic health of the United States; in 2006, the country exported $28 billion in aircraft and $16 billion in jet engines (Marder, 2007).

There are three primary systems on modern aircraft: the propulsion system, the structure, and the avionics (the electronics, computing, radar systems; Schafrik and Sprague, 2004a). The first two are discussed in terms of the historical perspective of material development, as a way of illustrating the minerals important for current and future aircraft performance.

The first patent on a jet engine propulsion system was filed in 1929, and in 1941 the first jet aircraft flew with an engine based on this patent, with routine operation of jet fighter aircraft by the Germans by the end of World War II (Schafrik and Sprague, 2004a). The improvement in jet engines over the years can be measured in terms of their thrust-to-weight ratio. The aircraft based on the original patent had a thrust-to-weight ratio of 1.5:1 compared to 6.8:1 in modern aircraft, and the industry desires to reach 10:1 in the future. These modern propulsion systems are possible only because of the improvement in the high-temperature properties of materials. The first jet engines were made of steels whose high-temperature properties placed severe limitations on operation. New alloys employing nickel, such as Inconel and Nimonic, greatly improved these properties (Schafrik and Sprague, 2004a). So-called superalloys based on nickel or cobalt had significantly improved mechanical properties at elevated temperatures. These improvements were based on additions of titanium and aluminum, which caused the formation of precipitates in the materials. Further improvements came from the introduction of titanium-base alloys (Schafrik and Sprague, 2004b).

Because of its much lower density compared to earlier alloy materials, titanium has become a preferred material for a number of gas turbine components, lending a significant reduction in weight. However, one of the drawbacks and continuing issues with respect to titanium is its higher cost (Schafrik and Sprague, 2004b). Titanium ore (which is reasonably abundant) is obtained primarily from Australia in the mineral rutile (TiO_2) or ilmenite (Marder, 2007). A chlorination process that leads first to what is termed titanium sponge produces titanium metal. The sponge is then crushed and made into pure particles, mixed with alloying elements, consolidated, and melted (Schafrik and Sprague, 2004b). The United States has some titanium processing capability, but for economic reasons, companies are increasingly relying on foreign sources of titanium metal (Marder, 2007). This factor affects not only the commercial aerospace but also the defense aerospace industry.

A development that has further enabled higher engine temperature operation, important for the extra power requirements during aircraft take-off, is the use of thermal barrier coatings on the metal alloy blades. These coatings, based on yttrium oxide-stabilized zirconium oxide, are applied by plasma spraying or by physical vapor deposition (Schafrik and Sprague, 2004c).

When examining the aircraft structure, an interesting point is that the first aircraft to fly, built by the Wright brothers in 1903, was constructed largely from wood and fabric, both of which are composite materials. Over the next 20 years aircraft were built mostly from wood and fabric; metals were used only for engines, bracing, controls, and landing gear (*http://www.au.af.mil/au/awc/awcgate/vistas/match3.pdf*).

However, enhanced performance required new materials that were based on high-strength steels and aluminum alloys. Creation of suitable alloys, optimization of their heat treatment techniques, inexperience with their fabrication, and their cost were the main reasons for developmental difficulties in the metal airplane structure (Nye, 1935). The development of a corrosion-resistant aluminum alloy, duralumin, was a major breakthrough in the use of metals in aircraft construction, and other aluminum alloys followed. Figure 2.1 illustrates the time line for the development of aircraft materials. Interestingly, only two new major structural materials,

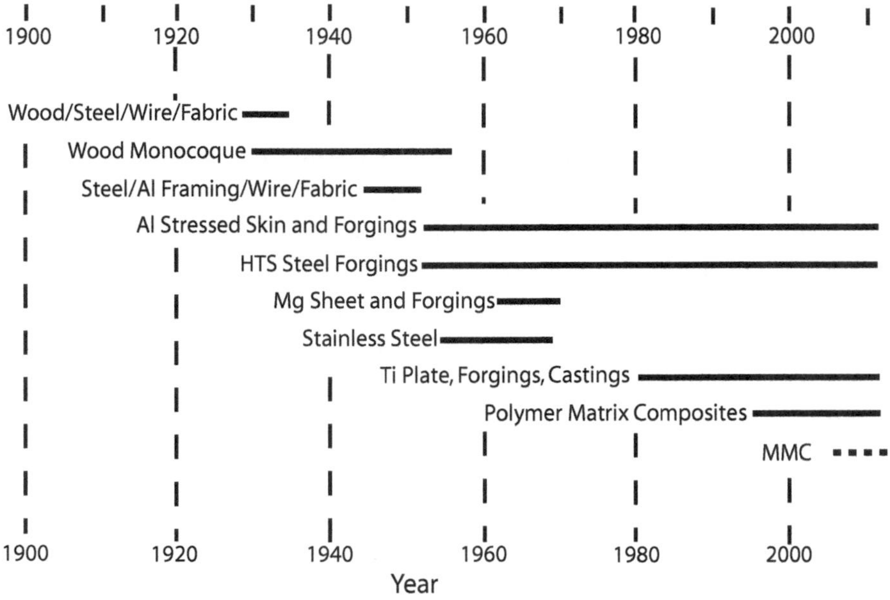

FIGURE 2.1 Aerospace frame materials and time of their introduction. Al: aluminum; HTS: high tensile steel; Mg: magnesium; Ti: titanium; MMC: metal matrix composites. SOURCE: Adapted from Air Force Scientific Advisory Board, 1996.

titanium and polymer matrix composites, have been introduced over the last 50 years.

Composite materials including both metal-matrix and polymer-matrix materials play an increasingly important role in aerospace construction. Reinforcements in these composites can consist of carbon or other fibers. As for automobile bodies, the future may well be the use of reinforcements consisting of carbon nanotubes having significantly enhanced stiffness and strength.

Electronics

The electronic age is here and the electronics sector, which includes computing, communication, entertainment, and dozens of other applications, is demonstrative of the dynamic nature of changing mineral and mineral suite applications that have facilitated technological advances. Miniaturization, energy efficiency, and increased processing or operating speed are some of the product performance goals that have driven research to optimize the properties of minerals or mineral products to meet new performance specifications.

On the basis of weight, steel and plastics are the two dominant materials used in electronic products. From the value distribution perspective, precious metals account for a significant part of the value for computers, cell phones, and calculators, television boards, and digital versatile disc (DVD) players. However, the electronic performance of the huge range of products in this sector relies on properties of elements derived from as many as 60 minerals (Hageluken, 2006). A large number of these elements are used as compounds formed with other solid or gaseous elements. These chemical compounds possess unique electrical, dielectric, or optical properties based on their atomic structure. In Chapter 1, the use of tantalum in cellular communications was mentioned (Box 1.1). Tantalum is employed as the complex compound barium-zinc-tantalum oxide ($Ba_3ZnTa_2O_9$) in this application; the compound has unique electronic properties that enable it to act as a resonator in the cellular telephone base stations. Substitutes for tantalum in this compound have proved ineffective. Another example in the electronics sector is indium, employed as a compound semiconductor, indium-gallium-arsenide (InGaAs), as well as an oxide in the manufacture of electroluminescent panels ("flat screens"). A third example is the use of hafnium as hafnium oxide in high-dielectric-constant films on silicon, important for the operation of microelectronic chips. The exciting new area of high-temperature superconductivity makes use of REs as well as other minerals to form complex compounds that possess unique electrical conductivity properties at temperatures far above those for superconducting metals (Hageluken, 2006).

Computer chip technology developments during the past 25 years (Figure 2.2) illustrate the increase in the number of minerals or their derivatives used in chip manufacturing as technological advances have captured increasingly more specialized mineral properties. The result is a dynamic view of mineral products through time and a view of the global perspective needed by computer chip manufacturers to secure the raw materials for chip production.

Energy

Energy production is currently receiving considerable attention, in part because of the increasing cost of fossil fuels, but also because of a growing desire to avoid generating greenhouse gases. While minerals are clearly important in current energy production—for example, in oil and gas extraction, refining, and mining of fossil fuels—the development of alternative energy sources will demand new materials and minerals. A few examples are presented below.

Solar electricity has experienced a global surge of popularity over the past 5 years as a source of renewable energy for private and industrial purposes. Photovoltaics (PVs; conversion of sunlight to electricity) are popular partly because of their adaptable power outputs, low environmental impact (noise- and pollution-free energy production), and low operating costs (Benner, 2007). Although PVs presently provide only 1 percent of domestic energy consumption from all renewable energy sources, and renewables themselves comprise only 6 percent of all sources of domestic energy con-

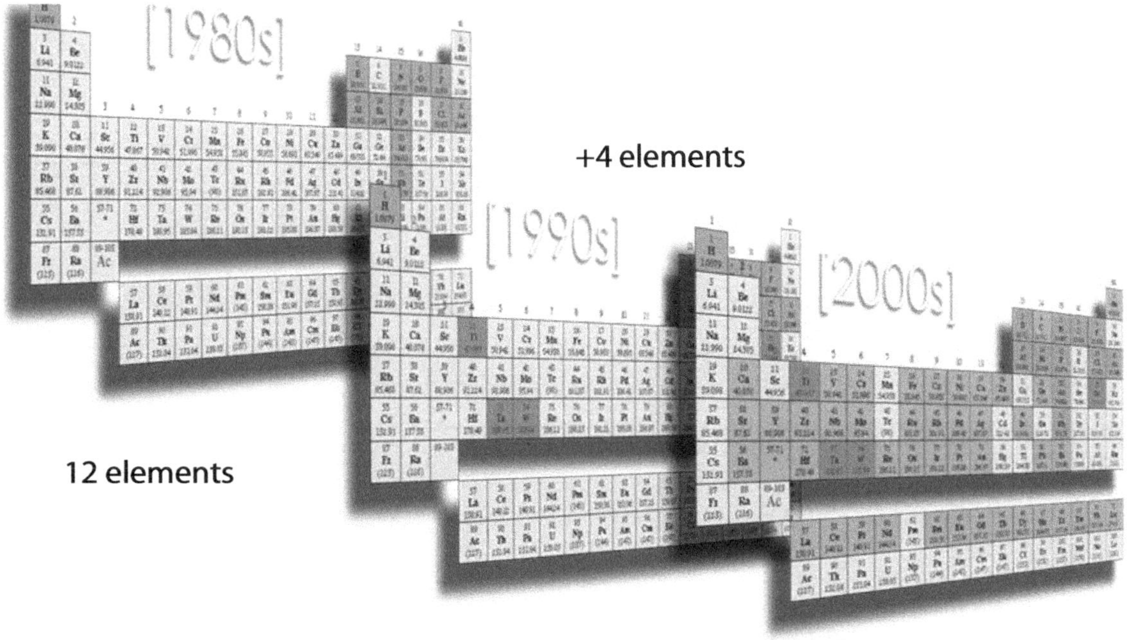

FIGURE 2.2 The dynamics of two decades of computer chip technology development and its mineral and element impacts. In the 1980s, computer chips were made with a palette of twelve minerals or their elemental components. A decade later, 16 elements were employed. Today, as many as 60 different minerals (or their constituent elements) may be used in fabricating the high-speed, high-capacity integrated circuits that are crucial to this technology.
SOURCE: Used with permission from Intel Corporation.

sumption (EIA, 2007), accelerated growth in solar systems as an electricity source is projected to take place over the next several decades given specific levels of national and state investment (Table 2.3). Global PV cell production has increased six times since 2000, and some projections suggest PVs will provide half of new U.S. electricity capacity by 2025 (Benner, 2007).

The material requirements for a PV installation include construction materials such as copper, steel, concrete, aluminum, and glass. However, the PV cell is the specialized feature of the system, with the majority of cells manufactured as silicon wafers; cadmium-telluride (CdTe) or copper-indium-gallium-selenide (CIGS: $Cu(In,Ga)Se_2$) thin film; and very high-performance cells made of gallium-arsenide (GaAs) compounds, for example, constitute the other cell types. To accommodate the anticipated increase in demand for PV cells, high-purity silicon production is slated to double by 2010. Better processing technology and advances in the use of thinner silicon films and wafers will likely preclude any disruption in the supply of domestic refined silicon for this application. Similarly, despite the spike in price of 9 to 15 times between 2002 and 2006 for indium and tellurium, respectively, the small quantity of these materials used in PV wafers has caused only a slight increase in the cost of module production until the present. The rapid growth in use of indium tin oxide to produce liquid crystal displays has partly fueled the price increase of indium in this 4-year period. Stable or falling prices for cadmium, selenium, and gallium over the same period have contributed positively to their use in the manufacture of PV wafers.

TABLE 2.3 U.S. Solar Power Growth Projections Through 2030

Solar Power Systems (gigawatts)	2004	2015	2030
Annual U.S. shipments	0.12	2.3	19
Cumulative U.S. installations	0.34	9.6	200

NOTE: U.S. figures are Roadmap Projections from Solar Energy Industries Association.
SOURCE: SEIA, 2004.

Price volatility for tellurium and indium may impede their more widespread use as a thin-film option for PVs (Benner, 2007), although indium, as a by-product of zinc processing, is always subject to swings in price that are not related directly to raw indium supply. Differences have been noted between official U.S. government and industry sources with regard to data on supply and occurrence of tellurium and indium (Benner, 2007) and highlight the need, particularly during times of volatile price swings, to maintain unbiased, independent public mineral data sources that are cross-referenced to available industry data reports.

Batteries are ubiquitous power sources for many hand-held appliances, information technology and telecommunication devices, computers, motor vehicles, and aerospace applications. Lithium ion and nickel-metal hydride rechargeable batteries are preferred in many applications largely because of their long life cycles and high energy densities. Recycling of lithium batteries is considered important with regard to this supply outlook. Recycling or disposing of lead-acid or nickel-cadmium batteries requires handling potentially toxic substances more often than would be the case with long life-cycle rechargeable lithium ion or nickel-metal hydride rechargeable batteries.

Nickel-metal hydride batteries have a slightly lower energy density compared to lithium ion batteries but are currently the most commonly used battery in hybrid electric vehicles. Nickel hydroxide forms the cathode of the battery, while the anode is composed of a metallic compound of REs (particularly lanthanum and cerium) and another metal (often cobalt, nickel, manganese, or aluminum) (*http://www.cobasys.com/pdf/tutorial/inside_nimh_battery_technology.pdf*). Estimates of about 20 kilograms (44 pounds) of REs are used in a hybrid car, between the rechargeable battery pack and the permanent magnet motor and regenerative braking system (*http://www.gwmg.ca/aree/faq.php*).

Lithium ion batteries have become a common substitute for nickel-cadmium batteries in many hand-held electronic devices, largely because of a high energy-to-weight ratio and slow charge loss when not in use (Olson et al., 2006). Although safety issues with lithium ion battery volatility have inhibited wider implementation of the batteries (Olson et al., 2006), they are viewed by some as a potential future substitute for nickel-metal hydride batteries (Herring, 2007), particularly in hybrid vehicles.

Some Minerals Decline in Use While Others Increase

Not all minerals have increased in use over time. Changes in technologies as well as concerns regarding the environment, health, and safety, combined with new government regulations, have resulted in a significant decrease in the use of certain mineral products including arsenic, lead, and mercury in specific applications (Matos, 2007). Examination of some of these minerals is also illustrative of the importance of substitutability in determining a mineral's criticality. Arsenic has historically been used in the United States as a pesticide and fungicide. This use in agriculture was essentially eliminated in 1993. Another important use has been as a wood preservative, but since 2003 its use in this arena has been supplanted by other materials. Although arsenic continues to be used in a number of nonferrous alloys, electronic components, and glasses (Figure 2.3), the protection of groundwater from arsenic is so important that even naturally occurring concentrations are considered unacceptable and require removal before water can be consumed by the human population.

Lead is another example of a material in which concerns over health issues have led to its decline in use in some applications in the United States, although the overall consumption of lead continues to grow (1.6 million metric tons in 2006 versus 1.3 million metric tons in 2003; USGS, 2007a). At one time lead was an important additive to gasoline. However, its use in gasoline interfered with the operation of catalytic converters, leading to its complete elimination in 1995 through federal legislation. The use of lead in paints and pigments has also been considerably reduced; in the United States, lead has been banned from house paints since 1977. There is now considerable international pressure to eliminate lead from solders as well. This latter case is a good example of how a critical mineral can seriously affect performance; considerable effort has been expended in finding suit-

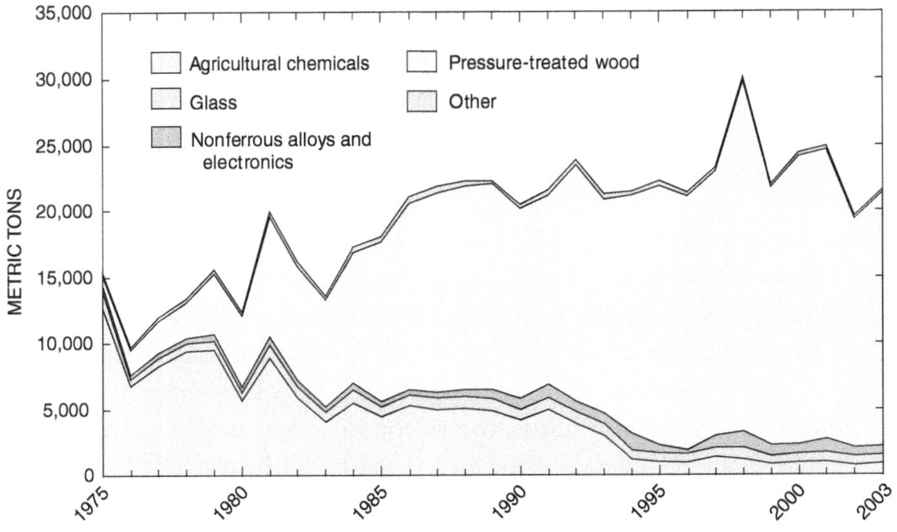

FIGURE 2.3 End uses of arsenic from 1975 through 2003. Rapid decline in use in the agricultural sector is clear after 1993. Since 2003, use of arsenic in pressure-treated wood has been curtailed. SOURCE: Matos, 2007.

able substitutes for lead-tin solders. Lead continues to be used extensively in lead-acid storage batteries, particularly for automobiles. In this application, the lead itself is recycled after the battery is spent.

Finally, until 1989, mercury was used extensively in nonrechargeable alkaline primary batteries. Although the amount of mercury per battery was small, the number of batteries sold made these the largest source of mercury in the municipal waste stream. Use of mercury in other applications has declined as well. Since 1992, mercury has not been mined as a principal product in the United States. Any mercury needs have been met through reclamation from fabricated products and as a by-product of gold production.

In contrast to these declines in demand, Matos (2007) notes a significant increase in the use of several other elements in specific applications:

- **Gallium** is used in integrated circuits, light-emitting diodes

(LEDs), photodetectors, and solar cells. Gallium arsenide has semiconducting properties and can convert electricity to light for use in LEDs. Gallium consumption in the United States rose 168 percent from 1975 to 2003.

- **Germanium**, used extensively because of its semiconducting properties in electronics, as well as optical glass fibers, is retrieved as a by-product of zinc and copper-zinc-lead ores, in which it is a trace element. Germanium also can be found in some coal deposits. An interesting new use of germanium is in chemotherapy for some types of cancer. Two companies in the United States produce pure germanium; the rest is imported from Belgium, China, and Russia. About one-fourth of the germanium consumed comes from recycling.
- **Indium**, of increasing importance for liquid crystal displays as the compound indium tin oxide, is a by-product of zinc ores. All of the indium in the United States is obtained from imports.
- **Strontium** is an element present in all television picture tubes in order to block X-ray emissions. However, as the use of flat screens, which do not require strontium, gradually increases, this use of strontium will decline.

IMPACTS ON THE U.S. ECONOMY

Minerals—along with air, water, energy, and other natural resources—are essential directly or indirectly to every sector of the economy. In purely economic terms, data from the U.S. Department of Commerce and the U.S. Geological Survey (USGS) suggest that the value added to the gross domestic product (GDP) by major industries that use processed mineral products in 2006 was $2.13 trillion compared to the total U.S. GDP of $13.2 trillion (Figure 2.4). So it is clear that the industries that use minerals and mineral products form a significant portion of the U.S. economy; however, it also should be noted that industries using any particular mineral or critical mineral represent a much smaller portion of the overall economy.

Related data for 2005 from the Bureau of Economic Analysis of the

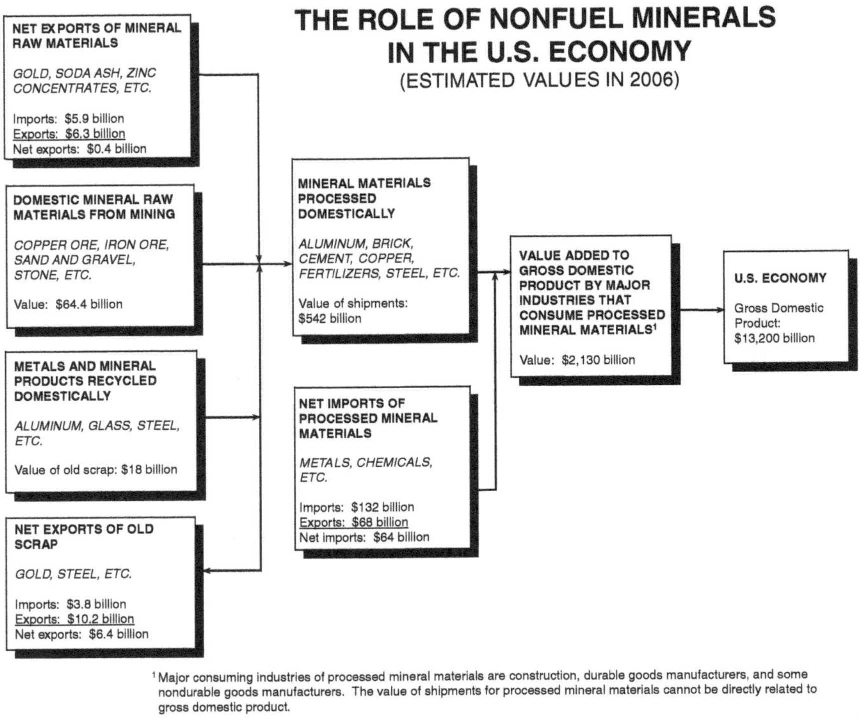

FIGURE 2.4 The role of nonfuel minerals in the U.S. economy. SOURCE: USGS, 2007a.

U.S. Department of Commerce (*http://www.bea.gov/industry/xls/GDP-byInd_VA_NAICS_1998-2006.xls*; accessed July 28, 2007) show how value is added to the U.S. economy prior to the actual use of minerals and mineral products in construction, manufacturing, and other sectors. Mining—excluding oil and gas extraction, but including mining support activities—constituted about $74 billion (~0.6 percent of total GDP) in value added; this compares to value added in oil and gas extraction of about $160 billion (1.3 percent of GDP). Durable goods manufacturing, including manufacture with nonmetallic mineral products, primary metals, and fabricated metal products, added $244 billion more to GDP (~2 percent of total GDP) from the mineral sector, whereas the comparable sector in energy (petroleum and coal products) constituted value added of $63 billion (~0.5 percent of GDP). A similar comparison to full- and part-time employment in these areas showed 128,000 persons employed in the oil and gas extraction sector and 437,000 employed in mining. An additional 1.5 million people were employed in the durable good manufacturing part of the mineral sector in 2005. Although it is difficult to estimate the partial, but important, direct fiscal or employment contributions of either petroleum products or minerals to the construction, plastics, transportation, automotive, aerospace, electronic, or general energy sectors, some estimates place that total contribution at approximately $2 trillion in value added to the GDP by industries that consume processed mineral materials (USGS, 2007a). These figures indicate that overall mineral use and associated employment are important to the U.S. economy. However, as discussed below, not all mineral use is subject to supply disruption; thus these figures ought not to be interpreted as indicators of how much of the U.S. economy is at risk should the supply of a specific mineral be restricted.

The information presented above illustrates the role of mineral production and use *overall* in the U.S. economy. Clearly a supply restriction for a specific critical mineral will affect only a part (typically a small part) of the overall U.S. economy. A supply restriction for any particular critical mineral will have a much smaller effect on the overall economy than would a restriction in the supply of oil, for example. Supply restrictions may come in one or a combination of two forms. First, demand may increase sharply in a short period of time; once producers are operating close to their production capacities, they find it difficult to increase output over the short term, even if demand exists for additional output (a demand shock). Second, on the supply side, supply may not be forthcoming for any one of a number of reasons described in Chapter 3 (a supply shock). In either case, if a mineral is not available to a user in a physical sense, the product that the user makes with the mineral or mineral product cannot be manufactured, sold, and then used by the purchaser of the product. In most cases, however, the consequences of a supply restriction of either form are more nuanced than simply whether a mineral or mineral product is available physically. Rather, a supply restriction represents an imbalance between supply and demand, which normally leads to higher prices, a reallocation of available supply toward those users who are willing to pay more for a mineral or mineral product and away from lower-valued uses, and finally an incentive for higher-cost sources of output to begin supplying the market. The magnitude of impact of a supply disruption on the domestic workforce and economy will depend on the ease or difficulty of (1) substituting for the mineral or mineral product that has become unavailable or more expensive and (2) quickly drawing on alternative, previously higher-cost sources of supply.

The specific impacts will clearly depend on the circumstances: Is the mineral physically unavailable? A physical shortage of a mineral is more likely to occur when the mineral is used in relatively small quantities in a limited number of applications (situations that Chapter 3 describes as *thin* markets). Have prices increased, which is more typical for minerals used in large quantities in many applications throughout the economy? If prices rise, by how much have they risen? How flexible or inflexible is demand (i.e., how easy or difficult is it to substitute for the restricted mineral)? How flexible or inflexible is supply (i.e., how easy or difficult is it to bring new sources of supply into production)? The types of possible effects include impacts on the following:

- *Domestic production of minerals:* opportunities may arise for increased domestic production of the mineral whose supply has been restricted, unless the supply restriction is caused by disruption at domestic production facilities; increased domestic production might incorporate previously uneconomic primary or secondary production.
- *Domestic users of minerals:* production might be lost due to lack of mineral availability and/or higher costs as a result of that lack of availability; in such a case, applications might be concentrated in higher-valued uses of minerals and entail higher-cost production; emerging use industries might experience slower growth; some industries might see lower profits.
- *Domestic employment:* domestic employment impacts might be reg-

istered in research and development, manufacturing, and engineering (Marder, 2007), as well as in extraction and processing, with the level of the impact depending on the scale and duration of the restriction.

SUMMARY AND FINDINGS

This chapter demonstrates that minerals are used throughout the economy. Some, such as iron and steel, are used in large quantities in a large number of end-use sectors. Others, such as the PGMs and REs, are used in much smaller quantities and often in a limited number of applications. In all cases, minerals are the material components whose chemical and physical properties are essential for performance in each application. In this sense, all minerals are critical. Minerals, in effect, compete with one another to provide material properties. Over time, substitution of one mineral for another has occurred as technologies evolve. Substitution is easier in some situations than others (e.g., substituting for arsenic, as opposed to finding a replacement for PGMs in catalytic converters). In this sense, the more difficult substitution is in a particular circumstance, the more critical the mineral is to the performance and success of the material or product.

Importance in use embodies the idea that some minerals are more important than others. Substitution is the key concept here. If substitution of one mineral for another in a product is technically easy or relatively inexpensive, for example, one can say that its importance is low; the cost or impact of a disruption in the supply of the mineral would be low. If, on the other hand, substitution is technically difficult or costly, its importance is high; the cost or impact of a disruption would be high. This concept of importance at a product level also includes the net benefits consumers receive from using a product—the benefits to human health of nutritional supplements or pollution-control equipment, the convenience of cell phones, the durability of an automobile, and so on. A mineral can thus be important at the level of a product, either technically or economically. A mineral also can be important at a scale larger than a product. A mineral might be important to the commercial success of a company and its profitability (importance at a company level). A mineral might be important in military equipment and national defense; production of a mineral—or products that use a mineral as an input—might be an important source of employment or income for a local community, a state, or the national economy (importance at a community, state, or national level). Again, the greater the cost or impact of a supply restriction, which depends greatly on the ease or difficulty of substitution, the more important is the mineral.

It is likely that U.S. dependence on a broad spectrum of minerals will continue to grow as more sophisticated and complex applications develop—for example, in the area of nanotechnology. Each application may not require large quantities of a given material but will put more stringent demands on purity, cost, and accessibility. With the growing realization of the potential impact of global warming, new technologies will be developed that mitigate the output of greenhouse gases. PVs, hybrid vehicles, and the rechargeable batteries to power them all require mineral inputs, often from sources that currently lie outside the United States. Continuing social and economic progress nationally and internationally in an environment of increasing global competition for financial, human, and natural resources requires the United States to maintain a competitive stance with regard to materials science and engineering; mineral information, analysis, and activity to supply the manufacture of these materials; and sustainable approaches to extraction, processing, and use of natural resources.

Specifically the committee has found the following:

- Most common products in the domestic economy owe their function and form to various minerals and mineral products;
- The importance of various minerals in these products has changed over time with changing technology;
- The "dynamism" of mineral importance through time means that mineral criticality at a given moment is a snapshot, rather than an enduring constant; and
- Direct domestic impacts on individual consumers, and the extractive and manufacturing sectors supplying those products, can be estimated if a restriction in supply of a critical mineral occurs.

REFERENCES

Air Force Scientific Advisory Board, 1996. New World Vistas: Air and Space Power for the 21st Century, Materials in the Current Air Force. Available online at *http://www.au.af.mil/au/awc/awcgate/vistas/match3.pdf* (accessed July 11, 2007).

Associated Press, 2007. Judge blocks mining in key Miami-Dade area, industry to appeal. *St. Augustine Record.* July 15. Available online at *http://www.staugustine.com/stories/071507/state_4715191.shtml* (accessed August 7, 2007).

Benner, J., 2007. Presentation to Committee on Critical Mineral Impacts on the U.S. Economy. Washington, D.C., March 7.

Chavasse, R., 2005. Developments in hybrid vehicles and their potential influence on minor metals. Minor Metal Trade Association and Metal Pages. Minor Metals 2005, Lisbon, Portugal. Available online at *http://www.evworld.com/library/hybrids_minormetals.pdf* (accessed October 29, 2007).

EIA (Energy Information Administration), 2007. Annual Energy Review 2006. Report No. DOE/EIA-0384(2006). Washington, D.C. Available online at *http://www.eia.doe.gov/emeu/aer/contents.html* (accessed June 27, 2007).

Gross, D., 1996. *Forbes Greatest Business Stories of All Time.* New York: John Wiley & Sons, Inc., 358 pp.

Hageluken, C., 2006. Improving metal returns and eco-efficiency in electronics recycling—a holistic approach for interface optimisation between pre-processing and integrated metals smelting and refining. Proceedings of the 2006 IEEE International Symposium on Electronics and the Environment. San Francisco, Calif., May 8-11, 2006. Available online at *http://ieeexplore.ieee.org/xpl/tocresult.jsp?isnumber=34595&isYear=2006&count=79&page=1&ResultStart=25* (accessed July 19, 2007).

Herring, I., 2007. Presentation to Committee on Critical Mineral Impacts on the U.S. Economy. Washington, D.C., March 7.

Kohler, S., 2002. Map Sheet 52, Aggregate Availability in California. Sacramento: California Department of Conservation, California Geological Survey.

Marder, J., 2007. Presentation to Committee on Critical Mineral Impacts on the U.S. Economy. Washington, D.C., March 7.

Matos, G.R., 2007. Effects of Regulation and Technology on End Uses of Nonfuel Mineral Commodities in the United States. U.S. Geological Survey Scientific Investigations Report 2006-5194. Available online at *http://pubs.usgs.gov/sir/2006/5194/pdf/sir20065194.pdf* (accessed October 29, 2007).

McCartan, L., W.D. Menzie, D.E. Morse, J.F. Papp, P.A. Plunkert, and P.-K. Tse, 2006. Effects of Chinese mineral strategies on the U.S. minerals industry. *Mining Engineering* (March):37-42.

MII (Mineral Information Institute), 2007a. Every American Born Will Need. Golden, Colo.: Mineral Information Institute. Available online at *http://www.mii.org/pdfs/2007_mii_Baby_Info.pdf* (accessed June 28, 2007).

MII, 2007b. Why Do We Mine? Golden, Colo.: Mineral Information Institute. Available online at *http://www.mii.org/pdfs/every/why1.pdf* (accessed May 17, 2007).

Morrell, S., 2006. The Initial Price Impact of Closing the Production of Crushed Rock in the Lake Belt Region of Miami-Dade County. Tallahassee: Florida Taxwatch. Available online at *http://floridataxwatch.org/resources/pdf/LakeBeltMiningImpact.pdf* (accessed July 19, 2007).

NRC (National Research Council), 2007. *Managing Materials for a 21st Century Military.* Washington D.C.: The National Academies Press, 144 pp.

Nye, W.L., 1935. *Metal Aircraft Design & Construction.* San Francisco, Calif.: Aviation Press, 227 pp.

Olson, J.B., N.E. Smith, C.R. Sheridan, and P.C. Lyman, 2006. Lithium-ion batteries for electric and hybrid electric vehicles. Proceedings of the 4th International Energy Conversion Engineering Conference and Exhibit (IECEC). San Diego, Calif.: American Institute of Aeronautics and Astronautics Inc., pp. 198-202.

Reynolds, J., 2007. China's hunger for African minerals. BBC News, International version. Available online at *http://news.bbc.co.uk/2/hi/asia-pacific/6264476.stm* (accessed July 2007).

Schafrik, R., and R. Sprague, 2004a. Saga of gas turbine materials, Part I. *Advanced Materials and Processes* (March):33-36.

Schafrik, R., and R. Sprague, 2004b. Saga of gas turbine materials, Part II. *Advanced Materials and Processes* (April):27-30.

Schafrik, R., and R. Sprague, 2004c. Saga of gas turbine materials, Part III. *Advanced Materials and Processes* (May):29-33.

SEIA (Solar Energy Industries Association), 2004. Our Solar Power Future: The U.S. Photovoltaics Industry Roadmap Through 2030 and Beyond. Washington, D.C. Available online at *http://www.seia.org/roadmap.pdf* (accessed September 13, 2007).

Sloter, L., 2007. Presentation to Committee on Critical Mineral Impacts on the U.S. Economy. Washington, D.C., March 8.

Stablum, A., 2007. Catch 22 for Hybrid Cars—Price vs. Volume. Reuters, March 27. Available online at *http://www.reuters.com/article/reutersEdge/idUSL2767808920070327* (accessed August 3, 2007).

USGS (U.S. Geological Survey), 2002. Rare Earth Elements—Critical Resources for High Technology. USGS Fact Sheet 087-02. Available online at *http://pubs.usgs.gov/fs/2002/fs087-02/* (accessed August 3, 2007).

USGS, 2007a. *Mineral Commodity Summaries 2007.* Reston, Va.: U.S. Geological Survey, 195 pp.

USGS, 2007b. *Mineral Industry Surveys Crushed Stone and Sand and Gravel in the Fourth Quarter 2006.* Reston, Va.: U.S. Geological Survey, 11 pp.

Wald, M.L., 2007. Costs surge for building power plants. *New York Times,* July 10, Section C, p. 1.

CHAPTER 3

Availability and Reliability of Supply

INTRODUCTION

The availability and reliability of the supply of mineral commodities relate to the horizontal axis of the criticality matrix described in Chapter 1. Availability is dynamic but is generally considered to be a long-term issue, whereas reliability of supply is a shorter-term issue. Part of the mineral resource endowment that is often overlooked is the amount of material that is landfilled or scrapped but could be recycled. Net imports and exports of scrap for recycling should also be taken into consideration. In addressing the availability of critical minerals and materials the availability of both the virgin resource (primary availability) and the previously processed resource (secondary availability) must be considered.

The committee defines five dimensions of primary availability in this study: geologic (does the mineral resource exist), technical (can we extract and process it?), environmental and social (can we produce it in environmentally and socially accepted ways?), political (how do governments influence availability through their policies and actions?), and economic (can we produce it at a cost users are willing or able to pay?). Geologic availability includes consideration of the geologically appropriate terrains for a given mineral, mineral associations, depths, grade, tonnage, and geometry of the deposit. Technical availability considers the state of technology and knowledge to find, extract, and process the mineral resource. Environmental and social availability includes attributes of the environment in which the mineral is found or processed, such as endangered species, water and air quality, and scenic beauty. Social availability accounts for the community acceptance of resource development and may be more commonly referred to as "social license to operate." Political availability applies at local, national, and international levels and is a function of the predictability of laws, the independence of the judiciary, the limits on litigation, the protection of land tenure, the willingness of the host country to allow or facilitate development of the resource and repatriation of profits, and the military and economic stability of a region and the availability of an appropriate workforce. Economic availability considers the cost to discover the mineral deposit; to extract the minerals; and to process, concentrate, and purify the minerals balanced against the market value of the product. The availability of technical and skilled workforces is also a factor in economic availability. This chapter discusses the dimensions of primary and secondary availability and additional indicators of risk to the supply to clarify the input used to evaluate the risk to the availability of a mineral as a determinant of that mineral's position in the criticality matrix.

THE FIVE DIMENSIONS OF PRIMARY AVAILABILITY

Geologic Availability

Mineral deposits often have specific associations with geologic terrains and vary in abundance as a function of geologic time; a few examples of minerals and their global geologic associations are listed here. The major source of copper from deposits, known as porphyry copper deposits, are most prevalent around the Pacific Ocean, along the west coasts of South and North America and in the South Pacific islands of Indonesia, and in Papua New Guinea (Figure 3.1). The deposits in the United States formed 50 million to 75 million years ago, while the deposits in the South Pacific can be as young as 1 million to 3 million years old. Porphyry copper deposits are low grade (0.3-1.0 percent copper) and large tonnage (often greater than 1 billion tons), with the copper-bearing minerals finely disseminated throughout the large volume of rock. Platinum group metal

(PGM)-bearing minerals (those containing platinum, palladium, osmium, iridium, or rhodium) tend to occur in narrow veins that can exist as part of layered igneous complexes. PGM deposits in the layered igneous complexes of the Bushveld Complex in South Africa or the Stillwater Complex in Montana are around 2 billion years old (Figure 3.1), while those of the Nor'ilsk-Talnakh district in Russia are between about 240 million and 260 million years old. Carbonatite deposits (calcium-rich igneous rocks), some of which host rare earth (RE) metals, can range in age from 1.9 billion years at Palabora, South Africa, to 1.2 billion years at Mountain Pass, California (Figure 3.1). Until recently, the two main mining locations for REs had been Bayan Obo in China and Mountain Pass in California; Mountain Pass was closed to active operation in 2002 (see Chapter 4). Carbonatite deposits such as Eden Lake, Manitoba, are also being explored for REs. There are hundreds of occurrences of RE-bearing mineralizations and several locations at which some RE metals could be produced as byproducts from other minerals with the right economic, technological, and regulatory conditions.

A common exploration approach is to look for mineral deposits in familiar terrain, in known geologic settings although new discoveries in unconventional areas are also made (Shanks, 1983). Additional research in an area of new mineral discoveries is completed to aid in the understanding of mineral and geologic controls on the deposit's distribution with the potential to lead to emergence of new mineral trends or a complete map of the extent of the initial discovery. For example, in the world-class gold belt of the Carlin Trend in Nevada, more than 180 million ounces of gold have been identified since the late 1960s. New discoveries continue to be made as our knowledge and understanding of the mineral deposits advances (Figure 3.2).

Even in districts or around mines that have been explored for a century, the knowledge of the geology can still be very incomplete and major discoveries of new resources have been made a hundred years after the original, due in part to improved mapping technologies and better understanding of mineral and ore systems. BHPBilliton discovered the 1.5 billion ton Resolution copper mineral deposit in 1995 underneath the Magma

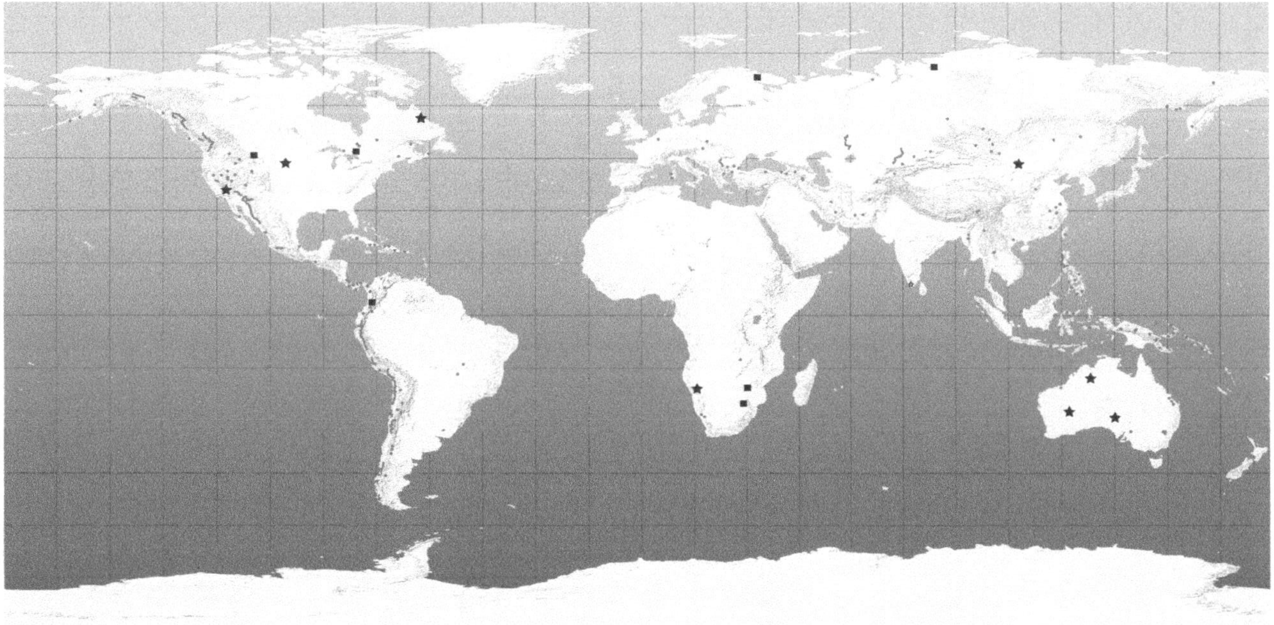

FIGURE 3.1 Distribution of porphyry copper deposits (red), PGM deposits (black squares), and magmatic RE deposits (black stars); RE deposits are only those reported to have greater than 1×10^6 metric tons of contained RE oxides. Deposit locations after Singer et al. (2005), Sutphin and Page (1986), and Jackson and Christiansen (1993). Numerous other types of sediment-hosted copper and placer-RE deposits also exist globally but are not plotted on this map. SOURCE: *http://veimages.gsfc.nasa.gov/2433/land_shallow_topo_2048.jpg*.

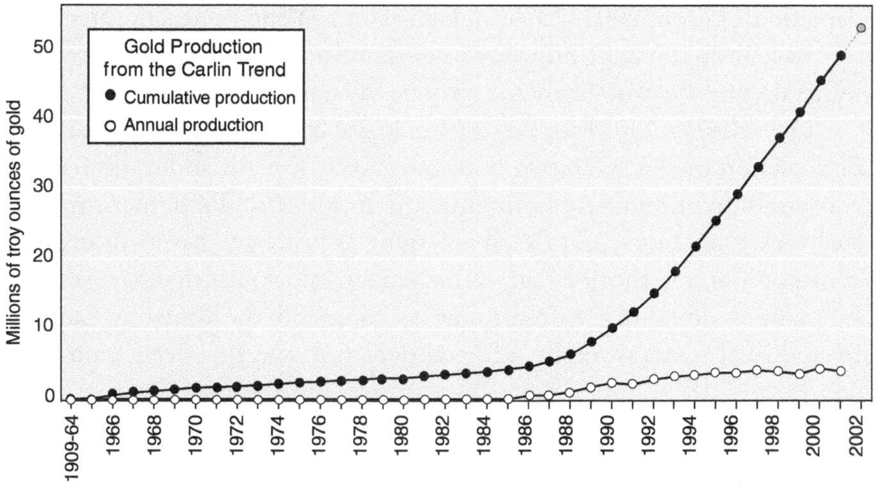

FIGURE 3.2 Annual and cumulative gold production from the Carlin Trend, Nevada. The trends from the late 1980s and early 1990s illustrate the effects on gold production of the discovery of deeper, sulfide-bearing gold deposits on the trend. SOURCE: Thompson and Teal, 2002.

ore body in Superior, Arizona. This area had been mined for copper since 1911 (Paul and Knight, 1995); however, the Resolution deposit remained undiscovered until exploitation methods and mining technology allowed more efficient and accurate exploration for deposits in unconventional areas at great depth. The Superior area has seen a resurgence of exploration activity, with additional copper resources being identified since the Resolution discovery. Thus, in addition to mineral exploration and discovery in new regions, new deposits may also be discovered in places where mines already exist.

Technical Availability

Mineral commodities can become more available over time if the cost-reducing effects of new technologies offset the cost-increasing effects of depletion (Tilton, 2003, 2006). Changes in mineral production technology have been dramatic and very important to the availability and cost of minerals, and they are likely to continue to be important (Box 3.1).

Over the last 130 years, new technologies have kept the adverse effects of depletion in check, despite both population growth and a surge in the consumption of mineral commodities (Tilton, 2006). With many nations such as China, India, and Brazil emerging as principal drivers of material consumption, and the price surges and reduced stockpiles that have resulted for many commodities, we must now question both the availability and the reliability of mineral supply. Mineral depletion and its effects tend to be

BOX 3.1
History of Advances in Mineral Production Technology

Mining and mineral processing have generally been at the forefront of industrial innovation for millennia and, through development of extraction and refining technologies, were responsible for major improvements in lifestyle, beginning with the Bronze Age and the Iron Age. Copper smelting, for example, began at least 4000 years ago. Until the mid-1800s, most aspects of mining and processing underwent progressive evolutionary improvements in technologies that had been applied successfully for centuries. However, the mineral industry then became a leading participant in the Industrial Revolution with important innovations in underground mining methods, improved gravity concentration equipment, and grinding mills.

Cyanidation of precious metal ores was commercialized in the early 1890s and led to rapid improvements in such areas as solid-liquid separation that soon spread to other industries such as waste water treatment. Open-pit mining was developed early in the twentieth century, enabling low-cost bulk mining of ore bodies with grades too low otherwise to support the costs of underground mining. Concurrently, the introduction of modern electric hoists made underground mining cheaper and safer. Selective froth flotation of metal sulfide ores quickly supplanted gravity concentration and significantly reduced processing costs and increased metal recoveries.

Through the remainder of the twentieth century, advances continued in all aspects of mining and processing, but with periodic lulls in the pace of innovation that were usually key drivers in increasing the costs of and prices for mineral commodities, although these increases may be mitigated in response to new technologies (Tilton, 2003).

The mineral resources of many mining districts or geologic regions are not known with certainty. As exploration technology advances and new geologic interpretations are produced, areas that were previously considered thoroughly explored are being revisited with new models and technologies at hand. For example, new drilling technology allows for deeper recovery of core and for holes to be drilled at subvertical angles. New analytical chemistry techniques allow more elements to be assayed at lower detection

caused by cyclical metal markets. A notable exception was the global gold industry, which was very active until World War II when War Production Board Order L-208 closed primary U.S. gold mines in October 1942. Little happened technologically in the gold industry until the 1970s when a gradual positive response to decontrolling of the gold price began to take place.

Since the 1960s, we have seen sweeping changes in production technology typified by the following brief list:

- Cheaper, safer, and more effective explosives;
- Bigger and more efficient excavators and haul trucks;
- Larger ore crushers with lower operating and maintenance costs;
- Cheaper and higher-capacity conveying systems;
- Heap leaching of low-grade gold ores;
- Treatment of cyanide solutions with activated carbon for gold recovery;
- A new generation of ultrafine grinding mills;
- Flotation cells that have increased in volume from 200 to 4000 cubic feet;
- Solvent extraction-electrowinning (sx/ew) for copper from leach solutions; and
- Flash smelting for metal sulfide concentrates.

These innovations and many others have enabled the mining industry to produce minerals, metals, and other elements at lower costs while making products of higher purity and greatly reducing the release of airborne and waterborne pollutants.

levels. The QEMScan™ technology, for instance, uses a sophisticated scanning electron microscope with four X-ray detectors and a microanalyzer to map bulk mineralogy, mineral textures, and metallurgical properties. New satellite data and imagery, including those from hyperspectral reflectance surveys, allow for more refined coverage of Earth's surface, contributing to better "remote" mapping of minerals. This type of technology assists in identifying regional mineral controls and trends, and zones of alteration that are prospective for certain types of minerals. Advances in many different geophysical techniques allow deeper exploration, higher resolution, or more accurate interpretations. One such advance is the Falcon™, the first airborne gravity gradiometry system, developed by BHPBilliton. Bell Geospace transferred submarine technology from the U.S. Navy to develop a full tensor airborne gravity gradiometry system. Both have provided significant advances in imaging potential mineral deposits at depth.

In 2002, the RAND Science and Technology Policy Institute published *New Forces at Work in Mining—Industry Views of Critical Technologies* (Peterson et al., 2002). The report said, "The United States has the largest mining industry in the world, with a raw material production of $52 billion in 1997. Yet many industry representatives noted that . . . mining is relatively small in comparison with other industries, and its ability to finance R&D [research and development] specific to mining is limited. As a result, many technology innovations in mining are adopted from other sectors such as construction, automobiles, and aerospace" (pp. 9-10). Technological advances are increasingly imported from countries such as Australia and Canada where public investment in mining-related research is at present greater than in the United States. The volatility of mineral commodity markets, the long delay in return on investment, and the unique requirements of mining equipment contribute to the financial risks for the mining industry and create difficulties for private companies to invest in research and development projects (Peterson et al., 2002). Tilton (2003) has suggested that, like exploration projects, a few highly successful research projects can more than compensate for the many less successful efforts. Opportunities for research and technology development in exploration, mining, in situ mining, and mineral processing are presented in the Peterson et al. (2002) report and ample discussion accompanies the recommendations.

Environmental and Social Availability

Objections to the development of mineral resources often focus on the disruption to the local environment and the impacts on communities related to the boom-and-bust nature of historic mining districts. Stories of the gold rushes in California and the Klondike and the resulting shifts in population, inflated prices, environmental damage, and social problems still resonate with the public. The growing development of the oil sands in northern Alberta, Canada, and the rapid growth in population in Fort McMurray, Alberta, highlight the issues that are faced when resource production expands faster than urban planning in an isolated community: housing may be in short supply, prices may become inflated, and the population may begin to feel torn between the improved economic prosperity and the disruption to the environment. Conflicts over land use in the rapidly urbanizing areas of the western United States often mean that the community must choose between the use of mineralized land for housing or recreation and its use for mineral resource development.

The Bureau of Land Management (BLM) and U.S. Forest Service administer 38 percent (393 million acres) of the land area in 12 western states, ranging from 76 percent of all land in Nevada to 23 percent in Washington State (NRC, 1999). In 1999, 0.06 percent of BLM land was affected by mining activity (current or planned) (NRC, 1999). However, not all public land is open to mineral entry and estimates from 1995 indicated that about 65 percent of western federal lands, or about 360 million acres, were restricted from mineral entry (Gerhard and Weeks, 1996). Since 1999, an additional 3.9 million acres have been withdrawn from mineral entry, and an additional 40 million acres have been proposed to be withdrawn. Wilderness areas are examined for their mineral inventory prior to withdrawal, but generally speaking, detailed mineral exploration is not conducted.

In some cases in the intermountain western United States, land with known ore deposits is effectively removed from mineral exploration by the development of surface rights for housing or other uses. Many western cities were located to take advantage of natural resources such as water, minerals, or timber. As these cities have grown to be major metropolitan centers over the last half century, conflict between development of natural resources and preservation or urban use of land containing the resources has sometimes occurred. Figure 3.3 shows eastern Maricopa and western Pinal Counties in Arizona, with 20 known copper deposits located within the area of the satellite image. The urbanization of the area has begun to overlap many discovered copper reserves such as the Poston Butte deposit near Florence, Arizona. The Poston Butte deposit was initially planned as

FIGURE 3.3 Satellite image of eastern Maricopa County and western Pinal County in Arizona covering an area with more than 20 discovered copper deposits. Urbanization has effectively removed copper reserves near Florence from mining development. Source: Barton, 2007. Used with permission.

an in situ leach operation of a deposit that contained 730 million tons at 0.38 percent copper (approximately 2 years of U.S. consumption), but when copper prices fell, the land was sold to a land developer.

From a mining company's perspective, the social availability of a mineral resource can be viewed as the need to obtain a license to operate. From a community's perspective, the goal of discussions about social availability is to break the boom-and-bust impact of mining on a community by developing a parallel economy and building independent capacity for development with power, water, transportation, communication, health care, and education infrastructure. Sustainable resource development is described by the Mining, Minerals, and Sustainable Development Project (MMSD, 2002) as the integration of economic activity with environmental integrity, social concerns, and effective governance systems.

Even with the implementation of sustainable development principles, a challenge for the mining industry is overcoming its often-negative legacy of distrust among some communities and stakeholders, and although the legal system may provide authorization for mineral exploration and development, social tension and conflict in a community can negate those rights. The relative rights of the local community versus the national community to benefit from the development of mineral resources are unresolved in many countries. The committee concurs with the MMSD in that the social license to operate at the local community level should ensure that "interactions between the mine and community should add to the physical, financial, human, and information resources—not detract from them" (MMSD, 2002, p. 198).

Sustainable development definitions abound and are best defined at the local level, integrating social, economic, environmental, and governance concerns with a basis in the local needs. No "one-size-fits-all" definition exists, and any definition must account for the unique needs at local scales. Sustainable development likely encompasses elements of all of the following: (1) the concept that the present generation behaves in a way that does not impede future generations from enjoying a standard of living at least comparable to its own; (2) the protection of an ecosystem, a community, an indigenous culture, and biodiversity; (3) assistance to communities that host mines to remain economically viable after the mineral resources have been mined; and (4) the equitable distribution of goods, income, and resources among the host communities and affected people.

Political Availability

The concept of environmental and social availability, the subject of the last section, leads naturally into, and in fact overlaps, the concept of political availability, which looks more directly at how actual government policies and actions influence mineral availability. The concept of political availability encompasses (1) legislation, rules, and regulations that influence investment in mineral exploration and mine development and (2) the risks and results of change in these policies. The former include rules about land access, security of tenure, permission to mine, and mineral taxation, as well as broader rules dealing with environmental and land access issues not designed with mining specifically in mind. The latter might be termed political risk—the likelihood of political decisions that alter the government rules under which mining occurs.

One of the most important issues related to development of the U.S. mineral endowment is the evolution of U.S. permitting regulations (NRC, 1999). Maintaining regulatory and leasing standards that keep pace with research on groundwater systems, soil chemistry, adaptability of plant and animal species, and engineering designs of the mine, for example, is difficult. Lack of planning and zoning in rapidly growing communities can result in differences of opinion over the juxtaposition of new residents and existing mines.

Environmental compliance and protection is not just a design consideration for new and existing mines but a necessity for operation from the time reconnaissance exploration is considered. The permitting process in the United States requires compliance with a body of laws administered through federal agencies and state and local authorities. Groundwater protection is one of the most cited reasons for protesting the development of a new mine. Hardrock mines and coal mines have a legacy of groundwater contamination, the most noticeable of which is acid-rock drainage in which the mineral pyrite oxidizes to form sulfuric acid, which lowers the pH of streams and causes iron to precipitate in stream sediments. Manganese, arsenic, uranium, and mercury may occur naturally in the groundwater but can be concentrated to potentially unacceptable levels by mining activity. Mining technology has advanced over the years to limit emissions into the environment as a result of mining activity in the United States. However, environmental protection regulations were established and have undergone subsequent revision within the last 30-40 years in most cases, and it remains important today to balance environmentally conscious mining practices and mine and environmental regulation. Mining companies do find it practical to work within the framework of today's environmental regulations, but permitting issues, overcoming the industry's past legacy of less environmentally friendly practices, and addressing community concerns in its development of new mines continue to offer time and resource challenges to the mining industry (Heig, 2007).

In addition to the legal framework for mineral development, the willingness of a nation to allow foreign commercial development of a resource affects its political availability. Nationalization of mines dates back to ancient Greece after the war between Athens and Sparta, and many similar examples have occurred around the world in the last 40 years. In this period of high commodity prices, resource nationalism tends to experience resurgence, raising the issue of supply disruptions in certain circumstances depending, for example, on government reinvestment in the mines, retention or departure of a technical mining workforce, and potential lapses in environmental and safety issues as mines change operational procedures and ownership.

The political challenges of mine nationalization inspired the founding of the Fraser Institute in 1974 as an independent research and education organization based in Vancouver, British Columbia. Since 1997 the Fraser Institute has published an annual survey on exploration and mining companies to evaluate how public policy influences exploration investment. The survey covers 65 countries and states on every continent but Antarctica and represents the opinions of exploration managers and company executives. The data reduction results in a metric called the Policy Potential Index

(PPI; see Figure 3.4). A perfect score of 100 means that a jurisdiction is perceived to be best in all measured aspects of public policy. A score of 0 means it is the worst in all categories. The Current Mineral Potential (CMP) index measures the mineral potential or endowment in a jurisdiction and whether the current policy environment encourages or discourages exploration. The Best Practices Mineral Potential Index (BPMI) measures whether policies conform to best practices. A room-for-improvement score is calculated by subtracting the BPMI score from the CMP. High positive scores on the room-for-improvement metric indicate that the jurisdiction is far from best practices. Details of the survey are in the report *Fraser Institute Annual Study of Mining Companies, 2006/2007* (McMahon and Melhem, 2007). While interpretation of the PPIs for a country (or a U.S. state or Canadian province) entails some subjective judgments and is somewhat dependent on available data, such a ranking is one means to try to quantify the rather qualitative term "political risk" with respect to mineral availability. Because it uses the same types of information over successive years to generate these evaluations, in a relative sense the PPI allows a company, government, or individual to gain some idea of the potential political risk entailed in mining exploration activities conducted in, or with, another country, state, or province.

Economic Availability

The manufacture of materials and products requires the availability of the raw mineral at a required time, quantity, and purity—at a price the manufacturer is willing and able to pay. Economic availability is the idea that a mineral is available at a price users are willing to pay and encompasses all of the previous dimensions of availability through the effects of these dimensions on costs.

Costs are fundamentally influenced by the abundance and availability of minerals, which are defined in terms of the level of detail of knowledge we have of a particular occurrence. A mineral occurrence is defined as an unusual concentration of a mineral or element that is of interest or value to someone (Cox and Singer, 1987). A mineral deposit is defined to mean that

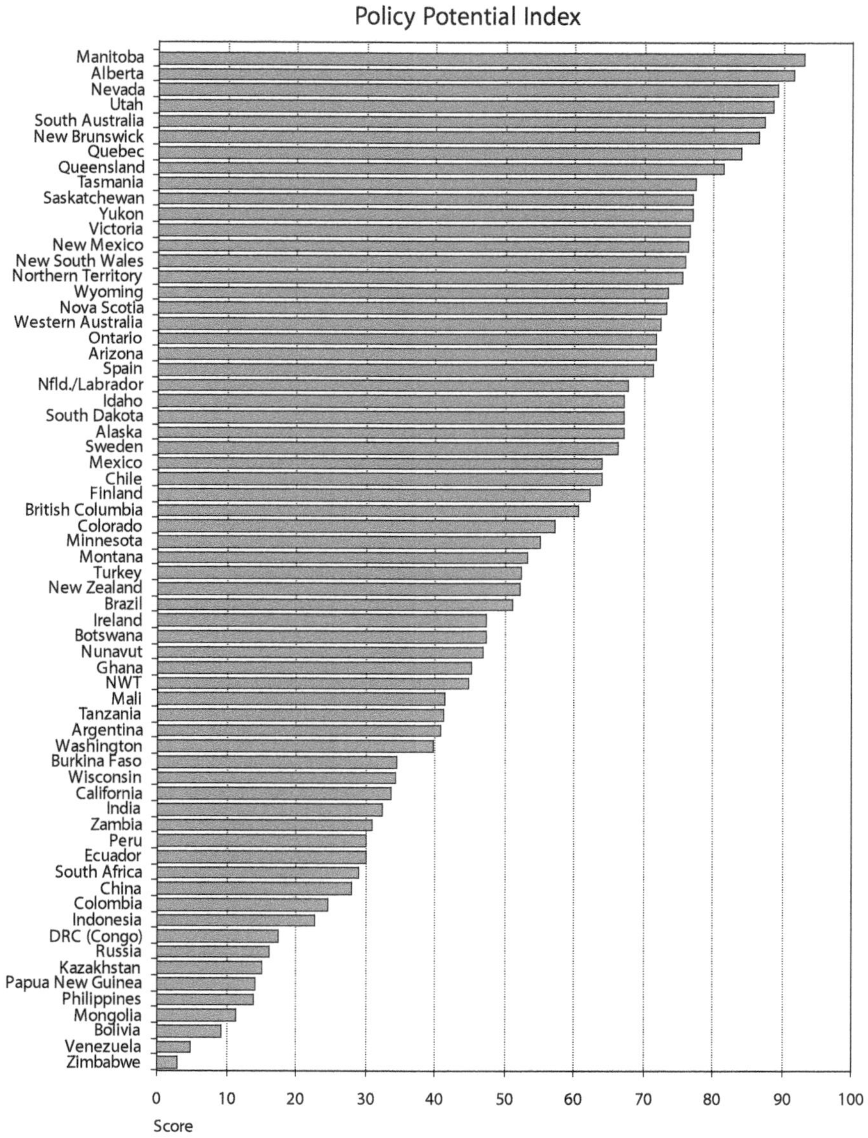

FIGURE 3.4 The relative rankings of jurisdictions for the Policy Potential Index. Source: McMahon and Melhem, 2007, p. 9. Used with permission.

the mineral concentration is worthy of further investigation as to economic grade and tonnage. An ore deposit is a mineral deposit of such grade (see Box 3.2), tonnage, or value that the minerals can be extracted, processed, and distributed at a profit. Mineral resources are categorized as measured (volume and tonnage well established), indicated (volume and tonnage estimated with less precise information), or inferred (deposit is assumed to occur past known boundaries), depending on the level of exposure of the deposit (Craig et al., 2001). A mineral reserve is a class of resource that is identified as economic as shown in Figure 3.5. A marginal reserve is "marginally economic." A change in the price of a commodity, exploration, extraction, or processing technology and changes in legal and regulatory policies can move a resource to a reserve and vice versa (Figure 3.6).

BOX 3.2
Grade

Two numbers often quoted with regard to resources and reserves are grade and tonnage. Tonnage is the amount of material containing the mineral commodity of interest. Grade is the relative quantity or percentage of the commodity or element of interest in a unit volume of mineralized rock. For example, in a porphyry copper deposit, the primary economic mineral may be chalcopyrite ($CuFeS_2$). The mineral contains copper, iron, and sulfur. By weight percent, about 35 percent of chalcopyrite is copper and the remainder is iron and sulfur. If 10 percent of the rock in the deposit contains chalcopyrite, 100 kilograms of chalcopyrite could be extracted from 1 metric ton (1000 kg) of mineralized rock. Since only 35 percent of the 100 kilograms is copper, one could expect, at best, to recover 35 kilograms of copper from that tonne of mined rock. This results in a grade of 35/1000 kilograms or 3.5 percent copper. A typical copper grade is between 0.5 and 1.0 percent which means that at a grade of 0.5 percent copper, we can extract a maximum of only 5 kilograms of copper from 1000 kilograms of mineralized rock. The actual amount of copper recovered is less because we are not able to extract all of the chalcopyrite from the rock, nor do we recover all of the copper from the chalcopyrite.

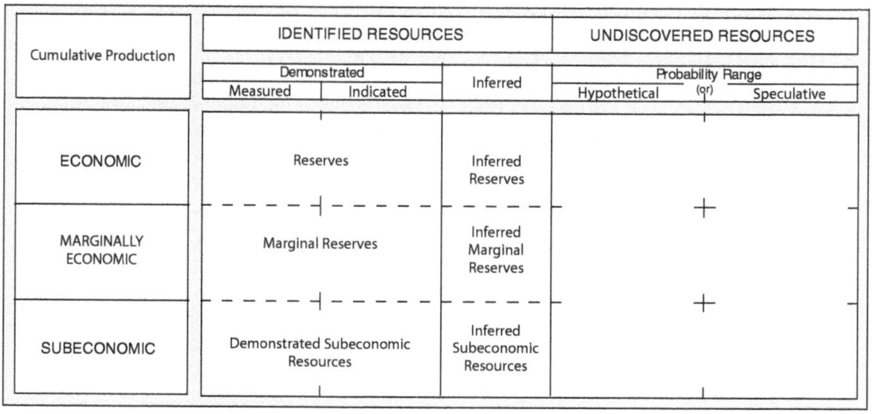

FIGURE 3.5 Classification of resources and reserves. SOURCE: USBM/USGS, 1980.

The definitions of mineral resources and reserves have legal implications for public companies filing financial statements. According to the Canadian Institute for Mining Standards, mineral reserves are "those parts of mineral resources which, after the application of all mining factors, result in an estimated tonnage and grade which, in the opinion of the Qualified Person(s) making the estimates, is the basis of an economically viable project after taking account of all relevant processing, metallurgical, economic, marketing, legal, environment, socio-economic and government factors. Mineral Reserves are inclusive of diluting material that will be mined in conjunction with the Mineral Reserves and delivered to the treatment plant or equivalent facility. The term 'Mineral Reserve' need not necessarily signify that extraction facilities are in place or operative or that all governmental approvals have been received. It does signify that there are reasonable expectations of such approvals" (CIM, 2005, p. 5). The U.S. Securities and Exchange Commission (SEC) defines a reserve as "that part of a mineral deposit which could be economically and legally extracted or produced at the time of the reserve determination" (SEC Industry Guides; available at *http://www.sec.gov/about/forms/industryguides.pdf*; p. 34).

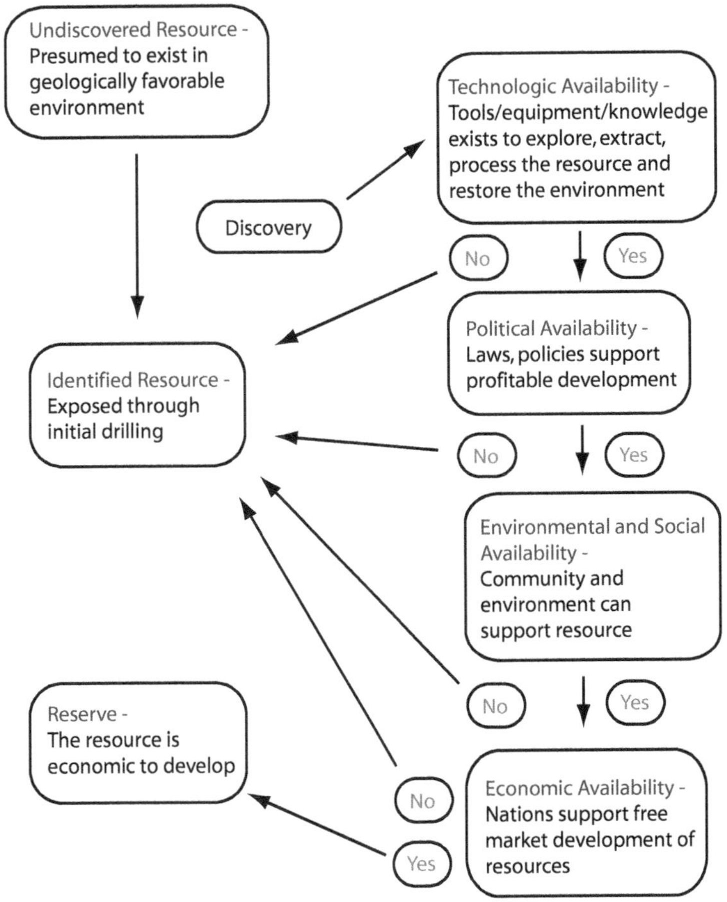

FIGURE 3.6 The availability of a mineral resource is dynamic in the five dimensions of geologic, technologic, environmental and social, political, and economic availability. Only if the extraction and processing of the resource is proved to be economically profitable is it considered a reserve.

A commodity is a physical substance such as grain, fuel, or a mineral that is interchangeable with a product of the same type and that investors buy and sell, often through futures contracts. There are risks associated with commodities because the producer does not know what the price will be in the future when the product is actually made and ready for sale. Mineral commodity prices are cyclical, but mining companies use long-term price predictions when planning new mines. "Price indexes developed by the U.S. Geological Survey (USGS) indicate that the long-term constant dollar price of key U.S. mineral raw materials declined over the last century, even though the need for mineral raw materials increased during the same period. Technologies and reduced production costs have allowed mineral production to remain profitable, while lower priced mineral products from domestic and foreign sources helped fuel growth in other sectors of the economy" (Sullivan et al., 2000, p. 2).

The USGS has analyzed the price trends of several mineral commodities using the Consumer Price Index for a constant-dollar comparison. The USGS calculates a U.S. mine price index using values for five metals (copper, zinc, lead, gold, and iron ore) and eight industrial minerals (cement, clay, crushed stone, lime, phosphate rock, salt, sand, and gravel). These commodities account for nearly 90 percent of U.S. metal and industrial mineral production. The overall price of mineral commodities declined in the twentieth century despite increases in consumption: supply and competition were adequate and technology improvements decreased the cost of production and increased the supply (Sullivan et al., 2000). Note that not all mineral commodities necessarily follow the downward trend in prices over time, although all of the industrial minerals included in the index exhibit a downward trend.

The implementation of bulk tonnage mining and processing methods at the turn of the twentieth century by Daniel Jackling allowed lower-grade porphyry copper deposits such as Bingham Canyon, Utah, to be developed. Jackling was an engineer working in the Bingham Canyon mines in 1899 who formed the Utah Copper Company, the predecessor of today's Kennecott Utah Copper Company. The use of large shovels and rail or truck haulage, coupled with froth flotation to produce copper concentrate, shepherded in the era of large open-pit metal mines with production costs that decreased as equipment size increased. The first commercially successful froth flotation process for sulfide minerals was the Potter-Delprat Process, which was first used (and not used much elsewhere because better processes soon appeared) at the Broken Hill Proprietary Mine in New South Wales, Australia, in 1903 (Taggart, 1927). Increasingly, new deposits will be discovered at depths that preclude surface mines. The twenty-first century will be characterized by the development of deep, high-production (100,000 metric tons per day) metal mines. Such mines have the benefit of limited surface impact and less visual surface impact. Although underground mines may be extremely expensive to develop and take 10-15 years of design and construction before any production occurs, they generally will have limited or no impact on water resources, because the dewatering requirements are restricted to smaller areas. Dewatering footprints will be reflective of the geometry of the aquifer as defined by structures, where present. The water in mineralized zones is often geologically and geographically separated from the water aquifers in valleys where drinking water is typically derived. Underground deposits are economic to develop when the long-term forecast for commodity prices is high and the developer can meet the economic requirements to move from resource to reserve. Technological challenges associated with equipment automation, ventilation, and hoisting still exist. The Resolution ore body underneath the old Magma Mine at Superior, Arizona, is one such example of a high-grade (1.5 percent copper) large-tonnage (1.5 billion metric tons) mine at 7000 feet (2134 meters) and rock temperatures of 180°F (82°C) (Heig, 2007). Examples of underground copper mines in the planning, construction, or production stages include Oyo Tolgoi, Mongolia; Grasberg, Indonesia; Pebble, Alaska; Superior, Arizona; Bingham Canyon, Utah; and Kidd Creek, Canada.

THE FOUR DIMENSIONS OF SECONDARY AVAILABILITY

The magnitude of the secondary resource in a particular region depends on past inflows and outflows from the stock of material available for recycling. Additions reflect historical use patterns within the region, product service lives, and imports of scrap for recycling, while removals reflect flows of material recycled within the region or exported to other regions for reuse or recycling. The secondary resource includes material discarded in landfills as well as material that is no longer in service but remains in place, material hoarded in anticipation of future shortages or price increases, and stockpiles of material awaiting reuse or recycling. The quantity and quality of the secondary resource is the result of many decisions, made by businesses, individuals, and governments over a very long period of time, that influence technical, economic, environmental and social, and/or political availability.[1]

The flow of material that becomes available for recycling in a particular time must be estimated to assess collection and recycling systems. The magnitude of this flow reflects more recent use patterns within the region and the distribution of product service lives. For each product or application it is necessary to determine the service life of products removed from service, the number of units sold in the year those products entered service, and the amount of material contained in each unit. Imports and exports of end-of-life products and scrap must also be considered. The flow of material available for recycling and the proportion that is actually recycled are the result of many decisions, made over an extended period of time, that influence technical, economic, and environmental and social availability. Decisions that affect product retirement and end-of-life management are especially important.

Technical Availability

Product use patterns and designs evolve over time. Some applications, by their nature, entail material losses over the life of the product that could potentially affect the quality of air, water, soil, or sediments. Examples include past use of tetraethyl lead in gasoline and a number of current uses, including the addition of micronutrients to fertilizers and animal feed supplements, use of zinc coatings to protect steel from corrosion, and

[1]This discussion of recycling and secondary resources refers almost exclusively to what is called "old scrap"—material in discarded and obsolete products that have reached the ends of their useful lives. The discussion of supply risk generally does *not* apply to scrap created during manufacturing, almost all of which is recycled soon after it is created and thus is not really a substitute for primary material. Sometimes known as "new scrap," this material is essentially primary material that requires an additional processing step to find its way into products.

use of inorganic copper compounds in friction products. Product stewardship and regulatory decisions require an understanding of the magnitude of losses and a comparative assessment of benefits and risks. Material that is dispersed will not be available for recycling in the future and will not increase the secondary resource. Past and current use patterns affect technical availability and must be taken into consideration in estimating the stock of secondary resources and the flow of material available for recycling (Box 3.3; e.g., Wilburn and Buckingham, 2006).

The service life for each application is a key determinant of the time that material becomes available for recycling and the extent to which societal needs could be satisfied through increased collection for recycling. Some applications such as beverage containers have comparatively short lives, and in principle, the demand for aluminum alloys, steel, and other materials used in beverage containers could be satisfied to a substantial degree by recycled material. In contrast, many applications of minerals, metals, and alloys in construction and infrastructure, transportation, and communications remain in service for decades. Even in a country such as the United States with a mature economy, well-established infrastructure, and modest growth rates, the quantity of material available for recycling may meet only a modest proportion of future demand. Material that is either not collected or not recovered augments the stock of secondary resources and increases the demand for primary material, unless total demand has declined significantly.

The range of opportunity for supplanting virgin ores with recycled material is controlled by the dynamic balance between the availability of the secondary materials for recycling and the total demand. If, for example, the rate of use of a mineral steadily increases over time and the material's in-use lifetime is finite, even complete recycling will be insufficient to meet continuously increasing demand for the mineral in various applications. If the rate of use of the mineral stabilizes, recycled material will be adequate to meet the demand if none is lost when discarded (a very unlikely scenario). If the rate of use of the mineral decreases, recycled material can, in principle, be the sole source needed to meet demand.

In most cases in the United States, mineral use has continued to increase over time. There are no known examples of saturation at the country

BOX 3.3
Pollution Control Technology and Motor Vehicle Fuels

A change in product use can profoundly affect demand for materials and availability of secondary resources. A mandate to reduce tailpipe emissions from vehicles to improve local and regional air quality drove the adoption of catalytic converters. This required elimination of tetraethyl lead use as an octane enhancer and antiknock compound. The change substantially reduced lead demand by eliminating its use in a dispersive application that represented the largest end use and had no potential for lead recovery. The major end use today is in lead-acid batteries, an application that does not result in material losses during the life of the product. A well-developed infrastructure for battery collection and recycling has improved end-of-life management. Industry efforts have been supported by federal and state government actions. Most states have adopted laws to ban disposal of lead-acid batteries in landfills which has provided additional groundwater protection. Collection, storage, transportation, and recycling are controlled by the Resource Conservation and Recovery Act, but a conditional exemption provides relief from certain requirements for shipments destined to a permitted facility. For 1999 through 2003, the recycling rate of lead available from lead-acid batteries in the United States was 99.2 percent. A typical new lead-acid battery now contains 60 to 80 percent recycled lead and plastic (BCI, 2005; Wilburn and Buckingham, 2006). The difference between these statistics reflects relatively slow growth in the domestic vehicle fleet and the size of each battery, as well as net exports of batteries and battery scrap.

Catalytic converters now represent the dominant use of platinum, palladium, and rhodium. Initially that demand could be satisfied only by primary material. As converters and vehicles reached the end of their lives, a new infrastructure was established to collect converters and recover these valued PGMs. Available data indicate that the United States leads other regions in recovery rates for PGMs from autocatalyst, but new autocatalyst demand (net of recycled PGMs) is the major use. Continued demand growth is anticipated, as discussed in Chapter 4.

level, although Müller et al. (2006) concluded that U.S. iron use saturated in about 1980 on a per capita basis. For a few toxic materials such as arsenic and mercury, use has decreased sharply over time, as discussed in Chapter 2 (see also Matos, 2007). Reasonably well-functioning recycling infrastructures exist for some materials and some product groups. For example, a substantial industry sector has developed to support the reuse and refurbishment of used auto parts and to recover materials from end-of-life vehicles. Among metallic materials, steel is produced and recycled in the greatest quantities. American steelmakers have invested to build electric arc furnace (EAF) capacity that accounted for 56 percent of domestic steel production in 2005 (Fenton, 2006), and those plants and downstream customers depend upon a steady supply of ferrous scrap from automobiles and other sources; about 90 percent of the feed to EAF facilities is from ferrous scrap (Steel Recycling Institute, 2005). Other enterprises have developed and applied a range of technologies to recover nonferrous metals from autoshredder residue, to add value by sorting particular aluminum alloys, and to recover metals and other materials from end-of-life electronic equipment.

Notwithstanding these examples, it is clear that technologies for materials recovery and recycling are much less well developed than those for processing primary materials. The Department of Energy and other agencies have supported the development and assessment of some promising technologies for sortation of aluminum alloys, among others. It is clear that although further technological development could enhance material recovery, perhaps the major obstacle to increased recovery from secondary sources is the lack of coherent policies and programs at the local, state, and national levels to increase waste diversion of end-of-life products for material recovery. For example, although recycled aluminum is used increasingly in other applications, recycling of used beverage containers has declined significantly due to lower collection rates (Das, 2006).

Environmental and Social Availability

Social attitudes toward resource conservation and recycling generally support more rather than less recycling, although these attitudes sometimes are not reflected in the behavior of individual consumers. For most types of postconsumer waste, consumers are the weakest link in the recycling value chain. Increased attention to public awareness and education could improve the economic and environmental performance of existing waste diversion programs and provide a more supportive environment to increase the recovery of materials and energy.

Political Availability

In principle, the United States has relatively few barriers that would preclude increased waste diversion for material recovery. However, federal government departments and agencies have not been given a mandate to lead a national effort, with the result that local municipalities and states have developed a patchwork of policies and programs. In practice, this uncoordinated approach makes it difficult for businesses to identify and assess investment opportunities and risks. For most types of postconsumer products, an efficient collection network and a small number of state-of-the-art recycling facilities would be sufficient to process all of the material available for recycling. An ad hoc approach that does not permit economies of scale or encourage collaboration at a national or regional level cannot maximize the economic, environmental, and other benefits of waste diversion for material recovery. In practice, the recovery and reprocessing of secondary material represent operational approaches that range from small, local facilities to large-capitalization, high-technology industrial facilities; note that the Institute of Scrap Recycling Industries represents 1200 different companies in the United States. An effort to increase the amount and efficiency of recycling in this country could benefit from a combination of government policies that encourage the existing industry to upgrade and integrate its facilities and networks.

Although the United States has not assigned a high priority to the recovery of secondary materials, other countries have taken a number of steps to address concerns about material availability and the reliability of supply. Because the United States has not ratified the Basel Convention on the Control of Transboundary Movements of Hazardous Wastes and Their Disposal (hereafter called the Basel Convention) and parties to the convention are prohibited from trading in hazardous wastes (as defined by the convention) with nonparties in the absence of an Article 11 agreement or arrangement, U.S. businesses may be unable to source sufficient feed to supply new or existing recycling facilities, particularly if domestic supplies decline as a result of changes in global markets and the value chains that supply emerging markets. The United States may thus lag behind the European Union, Japan, South Korea, and other countries as these nations strengthen their ability to access secondary resources, while the most rapidly growing economies secure access to primary resources.

Economic Availability

Unlike other types of materials, minerals and metals largely retain their chemical and physical properties and can generally be recycled repeatedly. While recycled metal may not meet technical specifications for all applications, it is priced in relation to virgin material and does not trade either at a significant discount or at a premium. Recyclable materials that contain minerals and metals are commodities with well-established commercial specifications and trade at prices that reflect both the current price of the recoverable components and the costs that must be incurred in order to recover them. For most metals and many mineral-based products, reduced transportation, processing, and energy requirements provide a significant cost advantage that could increase in the future if avoided greenhouse gas emissions can be monetized. Similar factors may increase the beneficial use of mineral-based materials in the future that are generally managed as waste at present, including slags from ferrous and nonferrous metal production and flyash from thermal power generating stations. Slags from new and old blast furnace ferrous metal production are used as a substitute for natural aggregates in a variety of applications, and new blast furnace slag is increasingly converted to a granulated cementitious material that is a substitute for portland cement. Slag from steel production has more limited uses as aggregates because of quality issues (USGS, 2007). Flyash from thermal power generation—graded according to its residual carbon content—is a substitute for portland cement in ready-mix concrete. Lower-quality flyash can be used as a stabilizer in highway construction applications. While power plant bottom ash is generally considered a waste product, it also is often used as a base product in highway construction (see also NRC, 2006). Minerals and metals represent a significant proportion of the residual value of many end-of-life products, such as motor vehicles, large appliances ("white goods"), commercial and residential heating, ventilating, and air conditioning equipment, and electrical or electronic products ("brown goods"). They provide an important economic driver for material recovery from end-of-life products that often contain a complex mix of materials.

Market forces provide a sufficient and ongoing driver for many resource conservation and recycling activities, generally within a well-established legislative and regulatory framework. Fabricators and manufacturers are motivated to increase material efficiency and minimize waste in order to improve profitability and reduce material input costs, waste disposal costs, and potential liability. Material that cannot be reused within the process, known as "prompt scrap" or "new scrap," generally retains significant value that can be recovered. Even when transportation and processing costs exceed the resulting revenue, firms will favor recycling if the net cost is less than the cost of disposal. Waste generators are required to identify and characterize wastes and to track waste volumes in order to satisfy legislative, regulatory, or other requirements, including corporate policy requirements, or to conform to environmental management system standards. Management systems and competitive pressures also drive continual improvement. For the most part, government interventions focus on high-risk areas, such as the treatment, storage, and disposal of hazardous waste.

The value of end-of-life products, or the cost to ensure their sound management, reflects the cost of collection, transportation, and initial processing (sortation). For the most part, physical processes are used to separate various types of scrap of merchantable quality (old scrap). In some cases, residual material with low or negative value is generated and must be sent for energy recovery or disposal in a landfill or incinerator, with the scrap processor paying transportation costs that may exceed any revenue. In some cases a residue may meet technical requirements for beneficial use (such as clay cover material for a landfill or energy recovery) and may be viewed as a product, but may represent a net cost to the generator because negotiated prices are influenced by the next best alternative for each party.

Recycling metals from postconsumer (municipal solid) waste generally is more costly than recycling materials from junked automobiles, demolished buildings, industrial machinery, and similar goods. The metal content of postconsumer waste is lower and more variable per unit of material that has to be processed. A market system, therefore, is less effective in dealing with postconsumer waste, if the objective is to maximize the amount of recycling that occurs. Citizens demand services and lower taxes from local authorities, while municipal waste managers and elected officials view waste diversion as an added cost, rather than an opportunity to avoid waste disposal costs and generate revenue. As a result, the metal content of recycled municipal solid waste varies widely, depending on programs put in place at a state or local level. With a range of approaches, there is an opportunity to examine the economic and environmental costs and benefits of alternative measures, including material recovery from unsorted municipal solid waste, source segregation by householders with curbside collection of recyclable materials, deposit-refund schemes for beverage containers, a variety of design alternatives for extended producer responsibility programs, and other models. In short, although recycling is already an important economic activity, there is a need to investigate whether more effective incentives and disincentives are necessary to increase recycling and reduce the rate of accumulation of secondary resources in landfills.

Economic availability can be reduced substantially when different materials are mixed. There is a need to carefully weigh collection and transportation cost savings that may result from combining different waste streams against the revenue losses and cost increases that result from additional handling, processing, and impurities. In some cases, existing infrastructure can be used with limited pre-processing. For example, white goods can be shredded together with automobiles, provided that ozone-depleting substances and components that contain polychlorinated biphenyls are removed first. In other cases, collection and transportation cost reductions may eliminate the potential for profitable recycling activities unless other funding is available to support responsible material management. Policy measures may also be needed to ensure an economic incentive for responsible recycling within the United States, including landfill bans, advanced disposal fees, export restrictions, or other measures.

SUPPLY RISK

The previous section discusses primary and secondary availability over the longer term. This discussion now considers more specifically the factors useful in assessing the degree of supply risk for a mineral in the short and medium terms from a national perspective, and in the context of the global trends in the sources and production status of minerals.

Short- and Medium-Term Factors for Supply Risk

In the short to medium term (periods of a few months to a few years, but no more than a decade), there may be significant restrictions of supply, leading either to physical unavailability of a mineral, or more likely, to higher prices—for a number of reasons. First, as noted previously, *demand may increase significantly and unexpectedly*, and if production already is occurring at close to capacity, then either a mineral may become physically unavailable or its price will rise significantly. Demand can increase more quickly than production capacity can respond.

Second, *relatively thin (or small) markets* are another indicator of possible supply risk. The key insight here is that small markets may find it difficult to increase production quickly if demand increases significantly. This issue could be important when evaluating supply risk for some so-called minor metals—such as gallium, tantalum, or vanadium—that at present have demand concentrated in a small number of applications but could experience rapid demand growth with development of a new application for the mineral or metal.

Third, supply may be prone to restriction if *production is concentrated.* If concentrated in a small number of mines, supply may be prone to restriction if unexpected technical or labor problems occur at a mine. If concentrated in the hands of a small number of producing countries, supply may be prone to restriction due to political decisions in the producing country.

The previous discussion in this chapter of political availability and growing resource nationalism is relevant here. If concentrated in the hands of a small number of companies, supply may be prone to restriction from opportunistic behavior by companies with market power. Market power may allow a powerful firm to raise prices opportunistically to take advantage of a weak buyer. The Herfindahl-Hirschman Index (HHI) provides a measure of market concentration or power and is used by the U.S. Department of Justice when investigating possible monopolistic behavior. This index is the sum of the squared market shares of all firms in a particular market—for example, an industry with three firms with market shares of 40, 40, and 20 percent would have an index of $40^2 + 40^2 + 20^2 = 3600$. Likely index scores range from about 1 to 10,000: the greater the concentration in a market, the higher the index number (and vice versa). The U.S. Department of Justice considers markets with index numbers between 1000 and 1800 to be moderately concentrated and those with numbers greater than 1800 to be concentrated. If a merger leads to an increase of more than 100 points in the index, the Department of Justice presumptively has concerns about possible anticompetitive consequences of the merger (U.S. Department of Justice, "Horizontal Merger Guidelines," available at *http://usdoj.gov/atr/public/guidelines/horiz_book/hmg1.html*; accessed June 21, 2007). Unfortunately, lack of sufficient data on company market shares made it impossible for the committee to calculate and evaluate HHIs for the minerals examined in this study.

Fourth, the supply of minerals that come significantly from *by-product production* may be fragile or risky. The key idea here is that the availability of a by-product is determined largely by availability of the main product (e.g., gallium as a by-product of bauxite mining). Thus, by-product production is relatively insensitive in the short term to changes in demand for the by-product. An increase in the demand for and, in turn, the price of a by-product may not result in significant additions to production capacity for the by-product. Likewise, a significant drop in demand for a by-product also may not result in significantly lower by-product production. As in the case of thin markets, minerals whose supply consists predominantly of by-products may not respond as quickly to demand increases as otherwise might occur. One exception would be a situation in which a significant amount of by-product mineral is not recovered at the time demand increases.

Finally, markets for which there is not significant *recovery of material from old scrap* may be more prone to supply risk than otherwise. As discussed earlier in this section, old scrap consists of discarded products, whereas new scrap is created during the manufacture of products. Recovery of material from old scrap influences supply risk in the following way: significant recycling of old scrap means that there is a pool of available old scrap from which material can be recovered. Part of this pool represents material in products discarded *this period*, and part represents material in products discarded *in the past* but not recycled previously. Material in the pool of old scrap exhibits a wide range of recycling costs; some material is of relatively uniform quality and is located close to recycling facilities, and thus has low costs; other material is of uneven quality, perhaps contaminated with other metals, is located at a distance from processing facilities, and thus has higher costs of recycling. As a result, recovery of material from scrap is particularly sensitive to price changes. When prices are high, it makes sense to recover material from the high-cost part of the pool of available scrap. When prices are low, much of the pool of available scrap remains unprocessed and is available for recycling later. In other words, the pool of available old scrap is an alternative source of supply should other sources become restricted and prices rise. The same argument does not apply to new scrap; almost all new scrap is recycled when or shortly after it is created because it tends to be of uniform quality, is not contaminated with other materials, is located close to reprocessing facilities, and thus tends to have very low costs of reprocessing.

There are two other possible indicators of supply risk, which are commonly cited and possibly useful—but only if interpreted with care. Both are commonly misinterpreted. The first is *import dependence*. The idea has been suggested that imported supply may be less secure than domestic supply. In fact, import reliance may be good for the U.S. economy, if an imported mineral has a lower cost and/or similar or better quality than an alternative domestic mineral. This is not to suggest that U.S. consumers should rely on foreign supplies if the source of the foreign cost advantage is a result of policies regarding environmental quality or worker health and safety that are below minimum international standards. At the same time, the United States needs to be cautious in imposing its environmental and labor standards on other countries; there may be good, local reasons for differences among countries in these standards. Thus import reliance is a potentially useful indicator but one that must be interpreted with care. Analysts must understand the definition. The USGS reports U.S. net import reliance as a percentage of U.S. consumption for a large number of minerals and metals. Net imports represent the physical quantities of imports less exports, adjusted for changes in inventories held by industry or government. In essentially all cases, dependence is measured either at the stage of mineral ore or concentrate or as refined metal. Thus, measured import reliance represents the dependence of mineral processors (in the case of ores and concentrates) or product manufacturers (in the case of refined metal)—and *not* the import dependence of final consumers. The perspective of the final consumer would have to include mineral quantities embodied in imported goods and exclude mineral quantities in exported goods.

One also needs to be cautious in interpreting actual estimates of import dependence. Just because measured import reliance is high does not necessarily imply that supply is at risk. In fact, in several situations, high measured import reliance may be no less risky than domestic supply if imports come from a diverse set of countries and firms or imported mineral or mineral product simply represents intracompany transfers within the vertical chain of a firm (e.g., imported concentrate to be smelted at a company's domestic smelter, imported refined metal to be transformed into a semifabricated shape or form at a domestic plant).

The second possible indicator of supply risk is the *reserve-to-production ratio*. As described earlier in this chapter, reserves are that portion of the Earth's stock of resource for a specific mineral that is known to exist and technically capable of being extracted at a profit under current market conditions. Dividing a mineral's reserves by current (annual) production gives a measure of how long reserves will last at current rates of production. The interpretation would be that the shorter the estimated lifetime of reserves, the greater is the supply risk. However, just as in the case of import dependence, this indicator of supply risk easily can be misinterpreted and must be used with care. As reserves become limited, firms have the incentive to explore for and develop additional reserves. Given that it costs time and money to develop reserves, firms do not fully explore and develop a mineral deposit at the time of initial development. Reserve development is an ongoing activity at mines, and mineral exploration for previously unknown mineral deposits is an ongoing activity as well. Moreover, technological innovation often makes it technically and economically feasible to extract minerals from what previously was geologically interesting but uneconomic rock—in effect, converting a mineral resource into a reserve. Changing economic conditions (prices and extraction costs) also continually influence what is—and what is not—a mineral reserve. With these qualifications, nevertheless, reserve-to-production ratios provide some useful insight into a mineral's availability and supply risk. A related measure is the *ratio of a mineral's reserve base to production*, which provides a similar but slightly longer-term view of a mineral's availability and supply risk. The USGS defines reserve base as the inplace demonstrated (measured plus indicated) resource from which reserves are estimated. The reserve base includes resources that are currently economic (reserves), marginally economic (marginal reserves), as well as some demonstrated subeconomic resources (USBM/USGS, 1980). The same caveats apply to this possible measure of supply reliability.

SUMMARY AND FINDINGS

This chapter has focused on the horizontal axis of the criticality matrix—the availability and reliability of mineral supply. The committee considered both primary and secondary supply in its assessment. The five dimensions of primary availability over the longer term (greater than about 10 years) include geologic (does the mineral resource exist?), technical (can we extract and process it?), environmental and social (can we produce it in environmentally and socially accepted ways?), political (how do governments influence availability through their policies and actions?), and economic (can we produce it at a cost users are willing or able to pay?). Secondary availability incorporates the same set of factors with the exception of geologic availability. Instead of virgin ore, secondary availability must rely on inflows and outflows from the stock of material available for recycling, which includes material discarded in landfills, material that is no longer in service but remains in place, material hoarded in anticipation of future shortages or price increases, and stockpiles of material awaiting reuse or recycling.

In addition to these longer-term factors, the short- to medium-term (a few months to no more than 10 years) risks to mineral supply include significant and unexpected increase in demand for a mineral; relatively thin (or small) markets; concentration of mineral production (in the hands of a small number of mines or producing countries); significant derivation of the mineral as a by-product (of the production of another mineral); lack of significant recovery from old scrap; import dependence; and a mineral's reserve base-to-production ratio.

Whether evaluation of the mineral supply risk is with respect to long-, medium-, or short-term interests, several of the availability factors often interact to varying degrees, and the associated data used to interpret these factors and their interactions require cautious analysis. The committee reaffirms the conclusion of the report *Mineral Resources and Sustainability: Challenges for Earth Scientists* (NRC, 1996) that the federal government should facilitate activities that sustain mineral supplies with respect to exploration, development, technology, and recycling because these may be longer-term issues to which the private sector and market forces alone are likely not sufficient to meet challenges of sustainability. Finally, efficient and environmentally conscious development of mineral supplies can be accomplished in a regulatory framework that is adaptive to change, including advances in technological capabilities and sound environmental and mining research.

With respect to the availability and reliability of mineral supply, the committee found the following:

- The uncertainties in knowledge of the nature of inferred mineral resources lead to uncertainty about the actual resource base for critical minerals.

- The stocks and flows of materials are inadequately characterized and difficult to determine, especially import and export as components of products and losses upon product discard (e.g., Wilburn and Buckingham, 2006). This lack of information impedes planning on many levels.
- Of the short- to medium-term supply risk factors, those most difficult to interpret are import dependence and a mineral's reserve base-to-production ratio; the data available to evaluate these factors are neither easily collected nor always quantifiable.
- Remanufacturing and recycling technology is a key component in increasing the rate and efficiency of material reuse, yet little research effort has been expended on developing this technology.

REFERENCES

Barton, P., 2007. Presentation to Committee on Critical Mineral Impacts on the U.S. Economy. Washington, D.C., March 7.

BCI (Battery Council International), 2005. National Recycling Rate Study: Chicago, Illinois, Smith, Bucklin, and Associates, Inc. Available online at *http://www.batterycouncil.org/BCIRecylingRateStudyReport.pdf* (accessed August 7, 2007).

CIM (Canadian Institute of Mining), 2005. CIM Definition Standards for Mineral Resources and Mineral Reserves: Prepared by the CIM Standing Committee on Reserve Definitions. Adopted by CIM Council on December 11, 2005. Available online at *http://www.cim.org/committees/StdsApprNov14.pdf* (accessed October 31, 2007).

Cox, D., and D. Singer, 1987. *Mineral Deposit Models*. USGS Bulletin 1693. Washington, D.C.: U.S. Government Printing Office. 379 pp.

Craig, J., D. Vaughan, and B. Skinner, 2001. *Resources of the Earth*. New York: Prentice Hall, 395 pp.

Das, S.K., 2006. Emerging trends in aluminum recycling: Reasons and responses. In T.J. Galloway (ed.), *Light Metals*. Warrendale, Pa.: The Minerals, Metals, and Materials Society, pp. 911-916. Available online at *http://www.secat.net/docs/resources/Emerging_Trends_in_Aluminum_Recycling06.pdf* (accessed September 2007).

Fenton, M.D., 2006. Iron and Steel Scrap; chapter in Minerals Yearbook 2005, U.S. Geological Survey. Available online at *http://minerals.usgs.gov/minerals/pubs/commodity/iron_&_steel_scrap/fescrmyb05.pdf* (accessed October 31, 2007).

Gerhard, L.C., and W. Weeks, 1996. Earth resources data: A basis for resource analysis and decision-making. *Environmental Geosciences* 3(4):76-82.

Heig, R., 2007. Presentation to Committee on Critical Mineral Impacts on the U.S. Economy. Washington, D.C., March 7.

Jackson, W.D., and G. Christiansen, 1993. International Strategic Minerals Inventory Summary Report—Rare-Earth Oxides. U.S. Geological Survey Circular 930-N, 68 pp.

Matos, G.R., 2007. Effects of Regulation and Technology on End Uses of Nonfuel Mineral Commodities in the United States. U.S. Geological Survey Scientific Investigations Report 2006-5194. Available online at *http://pubs.usgs.gov/sir/2006/5194/pdf/sir20065194.pdf* (accessed October 29, 2007).

McMahon, F., and A. Melhem, 2007. *Fraser Institute Annual Survey of Mining Companies, 2006/2007.* Vancouver: Fraser Institute, 91 pp.

MMSD (Mining, Minerals, and Sustainable Deveopment Project), 2002. *Breaking New Ground.* London: EarthScan Publications in association with the International Institute for Environment and Development, 441 pp.

Müller, D.B., T. Wang, B. Duval, and T.E. Graedel, 2006. Exploring the engine of anthropogenic iron cycles. *Proceedings of the National Academy of Sciences* 103(44):16111-16116.

NRC (National Research Council), 1996. *Mineral Resources and Sustainability Challenges for Earth Scientists.* Washington, D.C.: National Academy Press.

NRC, 1999. *Hardrock Mining on Federal Lands.* Washington, D.C.: National Academy Press, 247 pp.

NRC, 2006. *Managing Coal Combustion Residues in Mines.* Washington, D.C.: The National Academies Press, 256 pp.

Paul, A., and M. Knight, 1995. Replacement ores in the Magma Mine, Superior, Arizona. Pp. 366-372 in F. Pierce and J. Bolm (eds.), *Porphyry Copper Deposits of the North American Cordillera.* Tucson: Arizona Geological Society, 656 pp.

Peterson, D., T. LaTourrette, and J. Bartis, 2002. *New Forces at Work in Mining—Industry Views of Critical Technologies.* Santa Monica, Calif.: RAND Science and Policy Technology Institute.

Shanks, W., 1983. *Cameron Volume on Unconventional Mineral Deposits.* New York: Society of Economic Geologists, Society of Mining Engineers, American Institute of Mining, Metallurgical, and Petroleum Engineers, 246 pp.

Singer, D.A., V.I. Berger, and B.C. Moring, 2005. Porphyry Copper Deposits of the World: Database, Map, and Grade and Tonnage Models. U.S. Geological Survey Open File Report 2005-1060. Available online at *http://pubs.usgs.gov/of/2005/1060/of2005-1060.pdf* (accessed October 31, 2007).

Steel Recycling Institute, 2005. The Inherent Recycled Content of Today's Steel, Fact Sheet. Available online at *http://www.recycle-steel.org/PDFs/Inherent2005.pdf* (accessed August 8, 2007).

Sullivan, D., J. Sznopek, and L. Wagner, 2000. Twentieth Century U.S. Mineral Prices Decline in Constant Dollars. USGS Open File Report 00-389. Available online at *http://pubs.usgs.gov/of/2000/of00-389/of00-389.pdf* (accessed October 31, 2007).

Sutphin, D.M., and N.J. Page, 1986. International Strategic Minerals Inventory Summary Report—Platinum-Group Metals. U.S. Geological Survey Circular 930-E, 34 pp.

Taggart, A.F., 1927. *Handbook of Ore Dressing.* New York: John Wiley & Sons, 1679 pp.

Thompson, T.B., and L. Teal, 2002. Preface to *Gold Deposits of the Carlin Trend.* Reno: Nevada Bureau of Mines and Geology, p. 8.

Tilton, J.E., 2003. *On Borrowed Time? Assessing the Threat of Mineral Depletion*, Washington, D.C.: Resources for the Future.

Tilton, J.E., 2006. Depletion and the long-run availability of mineral commodities. In M.E. Doggett and J.R. Parry, (eds.), *Wealth Creation in the Minerals Industry: Integrating Science, Business and Education*. Littleton, Colo.: Society of Economic Geologists Special Publication 12, pp. 61-70.

USBM/USGS (U.S. Bureau of Mines and U.S. Geological Survey), 1980. *Principles of a resource/reserve classification for minerals: U.S. Geological Survey Circular 831*. Reston, Va.: U.S. Geological Survey, 5 pp.

USGS, 2007. *Mineral Commodity Summaries 2007*. Reston, Va.: U.S. Geological Survey, 195 pp.

Wilburn, D.R., and D.A. Buckingham, 2006. Apparent Consumption vs. Total Consumption—A Lead-Acid Battery Case Study. U.S. Geological Survey Scientific Investigations Report 2006-5155. Available online at *http://pubs.usgs.gov/sir/2006/5155/sir20065155.pdf* (accessed October 31, 2007).

CHAPTER 4

Applying the Matrix

USING THE MATRIX TO EVALUATE MINERAL CRITICALITY

The criticality matrix, introduced in Chapter 1, emphasizes that *importance in use* and *availability* (supply risk) are the key considerations in evaluating a mineral's criticality. This chapter evaluates the criticality of 11 minerals, selected on the basis of two considerations. First, the set of minerals the committee examined had to illustrate the range of circumstances that the matrix methodology accommodates and considers. For example, in the selection of the minerals examined, the committee considered minerals used in large quantities throughout the economy in traditional applications and others used in limited quantities in a small number of (often emerging) applications, minerals produced largely as by-products, and other minerals for which recycling of scrap is an important source of supply. Second, the set of minerals had to consist of those that, in the professional judgment of committee members, would likely be included in a more comprehensive assessment of all potentially critical minerals.

The next section examines in detail 3 of the 11 selected minerals or families of minerals: copper, platinum group metals (PGMs), and rare earth elements (REs). The section thereafter assesses in a more general manner the eight additional minerals that the committee considered potential candidates for criticality: gallium, indium, lithium, manganese, niobium, tantalum, titanium, and vanadium. The committee did not have the time or resources to provide a comprehensive assessment of all potentially critical minerals. The analysis rather focused on establishing the framework and criteria that might be considered by decision makers and mineral experts in determining a mineral's criticality and, subsequently, on assessing the type and frequency of information needed at a federal level to mitigate economic impacts should the mineral's supply become restricted. As a prelude to the criticality assessments, the committee reviews here the materials presented in Chapters 2 and 3 on mineral use and availability. These two chapters inform the "scoring" of the matrix for a specific mineral, or where in the matrix a specific mineral might fall at a given time.

The Vertical Axis: Importance in Use or Impact of Supply Restriction

The vertical axis, as noted previously, represents increasing importance in use, or analogously, the increasing impact of a supply restriction for a particular mineral. The methodology uses a relative scale of 1 (low) to 4 (high) to represent different degrees of importance or impact (Figure 4.1).

The key concept in locating a mineral on the vertical axis is substitution—the ease or difficulty of substituting for a mineral that becomes unavailable or too expensive. The position of a mineral on this axis depends on the context, the definition of which considers two important aspects. The first is *scale*. Are we concerned about a particular product and the impact a supply restriction would have on the performance of a product? Are we concerned about the effects on a local, regional, or national economy should the supply of a mineral essential to a local, regional, or national industry become restricted? Are we concerned about the effect of a supply restriction on a national priority, such as defense? For example, a mineral that is essential to the performance of a product (i.e., no ready substitutes exist to provide the same or similar performance) would be scored as a 4 by the manufacturer of the product. However, if this industrial sector was only a very small part of the national economy, it might be scored a 1 from the perspective of the U.S. economy. The second aspect of the context in placing a mineral on the vertical axis is *time*. The longer the period of time a user has to adjust to a supply restriction, typically the smaller is the consequence (substitution becomes easier). With a sufficiently long adjustment period, scientists and engineers usually can identify or develop a substitute

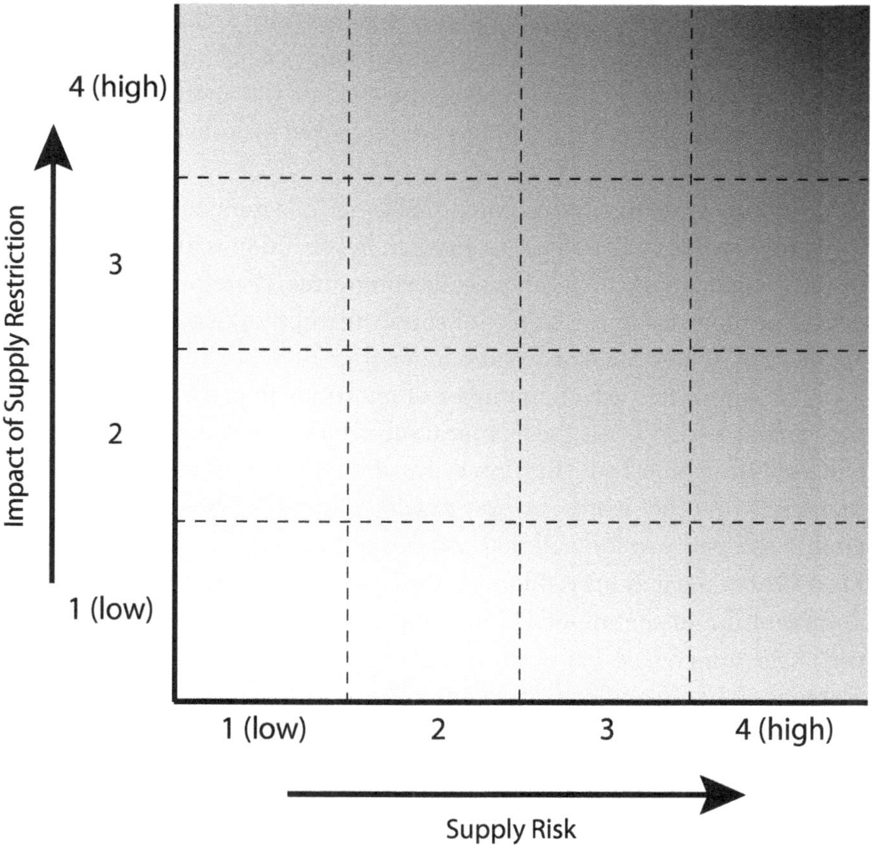

FIGURE 4.1 Criticality matrix diagram showing the two main factors that determine the scoring of a mineral's criticality: the impact of supply restriction (importance in use and ability to substitute for the mineral), and the supply risk (potential factors affecting the availability of the mineral). The axis scales are guides for the purpose of developing a weighted score for a mineral in terms of its criticality.

material with satisfactory or perhaps even better chemical and physical properties than the material whose supply was disrupted.

In the analysis that follows, the committee took a predominantly national perspective and one that is short to medium term (an adjustment period of one or several years, and no more than a decade). This framing of the analysis would not preclude someone else from evaluating a mineral from a local or regional perspective or over longer adjustment periods. As long as the context is established at the start of the analysis, the matrix concept can ideally be applied by anyone interested in evaluating a mineral's criticality.

It is important to recognize that the degree of a mineral's importance is likely to vary from one end use to another. Substitution is likely to be easier in some applications than others. The committee, therefore, evaluates the degree of importance (or impact of supply disruption) for each important application or end use of a specific mineral.

The committee asked a number of questions in placing a mineral on the vertical axis. What is the technical substitution potential in a particular end use? If technical substitution is possible, what are the economic consequences (in other words, by how much will production costs rise)? How vital is the end use for national considerations (e.g., national security)? How vital to the nation's economy is the industrial sector encompassing the dominant use of the mineral? How important to society is the dominant use of the mineral? What portion of the mineral will be used in emerging technologies or in applications expected to experience substantial growth? In the end, the actual placement of a mineral on the vertical axis represents the judgment of the committee considering these questions rather than the result of quantitative analytical assessment. Nonetheless, as shown below, the scoring is semiquantitative in that it attempts to weight the various application sectors for the mineral against the risk to the mineral's availability.

Finally, to facilitate consistency from one application of the matrix to another, the committee presents three indicators for each mineral it assesses: (1) estimated value of U.S. consumption of the mineral, giving an indication of the economic size of the sector (large or small); (2) the percentage of U.S. consumption in existing uses for which substitution is difficult or impossible (measured as the percentage of consumption; a score of 4 in this analysis indicates a high degree of impact from a supply disruption); and (3) the committee's professional judgment about the importance of growth in emerging uses that could overwhelm existing raw material production capacity in the short term (Table 4.1).

The Horizontal Axis: Availability and Supply Risk

The horizontal axis of the criticality matrix represents increasing risk (or probability) of supply disruption, which could exhibit itself not only in the form of physical unavailability of a mineral input but even more likely in the form of sharply higher prices for the mineral. As with the vertical axis of the matrix (importance in use), the way in which a specific mineral is placed on the horizontal axis depends on context. For supply risk, *time* is the essential aspect on which to focus. Are we concerned about availability in the longer term, over periods of a decade or more? Alternatively, are we concerned about the likelihood of short-term disruptions lasting weeks, months, or a few years?

In either case, analysis depends on considering the five fundamental determinants of a mineral's availability: geologic, technical, environmental and social, political, and economic. How these determinants are assessed depends on whether the analysis is short term or long term. For purposes of this chapter, as noted in the previous section, the committee primarily assesses short- to medium-term supply risks, while commenting on longer-term issues as appropriate.

When locating a mineral on the vertical axis, it is important to evaluate each significant application of a mineral separately because the degree of importance (ease or difficulty of substitution) typically varies from one application to another. When it comes to supply risk, however, the committee does not attempt to estimate different degrees of supply risk for different applications. At one level, especially when markets are large and well functioning, the supply risk is the same for all users. A supply disruption typically will exhibit itself in the form of higher prices that all end users face. The committee realizes that there may be circumstances in which different end use sectors face different supply risks. Such a situation might occur, for example, when one or a few large and powerful buyers are able to

TABLE 4.1 Criticality Indicators for Selected Minerals and Metals

	Copper	PGMs	REs	Niobium	Gallium
Relevant for Vertical Axis					
U.S. consumption (million $, 2006)[a]	16,625	1832	>1000	173	10
Percent U.S. consumption in existing uses for which substitution is difficult or impossible (4 in matrix)	15	55-90 depending on which PGM considered	44	32	~40 (indium-dependent)
Importance of growth in emerging uses that could overwhelm existing global production capacity (1 = low; 4 = high)	1	2	3	3	3
Relevant for Horizontal Axis					
Percent U.S. import dependence (2006)[b]	40	95 (Pt) 82 (Pd)	100	100	99
Ratio of world reserves[c]-to-production	31	139	715	73	NA
Ratio of world reserve base[d]-to-production	61	156	1220	87	NA

Indium	Lithium	Manganese	Tantalum	Titanium Mineral Concentrates	Titanium Metal	Vanadium
107	Not estimated	314	164	Not estimated	3255	68
~10 (partly gallium-dependent)	0	90	90	10 (for pigments)	90	11
3	2	1	2	1	2	2
100	>50 percent	100	87	71	Net exporter	100
6	194	40	33	122	NA	208
13	521	473	116	241	NA	609

continued

TABLE 4.1 Continued

	Copper	PGMs	REs	Niobium	Gallium
World by-product production[e] as percent of total world primary production	Small	Primarily coproducts	Primarily coproducts	NA	~100
U.S. secondary production[f] from old scrap, as percent of U.S. apparent consumption	7	Significant	Small	~20	0

NOTES: NA = not available.

[a]Estimated either as (1) the value cited in U.S. Geological Survey (USGS) *Mineral Commodity Summaries*, or (2) the product of U.S. consumption and price.

[b]Net import reliance as a percentage of apparent consumption. Net import reliance is defined as imports minus exports plus adjustments for changes in government and industry stocks.

[c]Defined by the USGS (2007) as "that part of the reserve base which could be economically extracted or produced at the time of determination. The term does not signify that extraction facilities are in place and operative."

obtain supply preferentially, even while other, less powerful end users are unable to buy a mineral at all or must pay a sharply higher price.

As with the vertical axis, the actual placement of a mineral on the horizontal axis represents the judgment of the committee, rather than the result of a quantitative analytical method. To assist in this evaluation, five indicators for each mineral that relate to current or future supply were also assessed (Table 4.1): (1) U.S. import dependence, which provides a starting

Indium	Lithium	Manganese	Tantalum	Titanium Mineral Concentrates	Titanium Metal	Vanadium
Most	Nil	Nil	Nil	Small	NA	Most
Small	Insignificant	Negligible	13	NA	<1 percent	Small

[d]Defined by the USGS (2007) as "that part of an identified resource that meets specified minimum physical and chemical criteria related to current mining and production practices, including those of grade, quality, thickness, and depth" and "the reserve base includes those resources that are currently economic (reserves), marginally economic (marginal reserves), and some of those are currently subeconomic (subeconomic resources)."

[e]Judgment based on published descriptions of production.

[f]Numerical estimates when available; judgment otherwise based on published descriptions.

SOURCES: Johnson Matthey, 2007; USGS, 2007.

point for evaluating short-term political risks, although one that is subject to many caveats; (2) the worldwide ratio of reserves to current production, giving an estimate of the lifetime of reserves; (3) the ratio of worldwide reserve base to current production, providing a longer-term perspective on geologic availability; (4) the relative importance of world by-product production in world primary production; and (5) the relative importance of U.S. secondary production from old scrap in overall U.S. consumption.

U.S. imports support investment in the exporting countries and generate social and economic benefits there. A high degree of import dependence for certain minerals is not, in itself, a cause for concern. Increased trade and investment flows contribute to economic growth and prosperity, both for the United States and for its trading partners. However, import dependence can expose a range of U.S. industries to political, economic, and other risks that vary according to the particular situation, including the country or countries concerned, the structure of the industry, and other factors.

The world reserve-to-production ratio integrates certain aspects of geologic, technical, and economic availability and is expressed in years. This term does not signify that extraction facilities are in place and operative. The world reserve base-to-production ratio is also expressed in years and integrates aspects of geologic, technical, and economic availability, but with less restrictive economic constraints. The ratio represents that part of an identified resource that meets certain physical and chemical criteria but includes resources that are currently economic (reserves), marginally economic (marginal reserves), and currently uneconomic (subeconomic resources). These two ratios represent high-level assessments based on an inventory of identified resources and key assumptions. Neither is based on a detailed, site-specific analysis of technical and economic feasibility. Classification of reserves and resources does not necessarily correspond to definitions used by regulatory agencies and relied on by investors. The ratios therefore provide an indication of the long-term availability of a mineral from primary sources. It is difficult to anticipate future exploration success, prices, costs, exchange rates, or production levels, all of which affect the resulting ratios. The underlying database may provide insight into the economic outlook for existing mines, but the ratios provide little insight into market dynamics.

Other sources of data, information, and analysis are required to assess the outlook for supply, demand, inventories, and prices of minerals over the short to medium term. Because such analyses are based on actual investment intentions and project evaluation activity, they provide a clearer indication of short- to medium-term availability, based on recent technical and economic assessments that take technical, political, economic, and other risks into consideration.[1] Project proponents and investors attempt to identify and mitigate project risks and secure insurance coverage against residual risks that cannot otherwise be controlled, but supply restrictions can arise from technical, environmental and social, political, economic, or other disruptions that were unforeseen, were unforeseeable, or could not be mitigated effectively.

Overall Assessment

The overall placement of a mineral on both the vertical and the horizontal dimensions of the matrix thus defines the degree of criticality of the mineral. The most-critical minerals are both essential in use (difficult to substitute for) and prone to supply restrictions. In the committee's view, criticality is best regarded as a continuum of possible degrees. There might, however, be specific situations in which a company or government agency would desire to create a list of "critical" minerals for the purpose of undertaking specific actions or policies to ensure supply or facilitate substitution away from highly critical minerals. The exact definition of what is critical (and by implication what is not critical) would depend on the specific context. Conceptually, however, a list of critical minerals would contain those minerals in one or more of the boxes in the upper right-hand portion of the matrix (Figure 4.1).

CRITICALITY ASSESSMENTS

This section applies the criticality matrix to three minerals or families of minerals: copper, REs, and PGMs. The committee selected these minerals because they exhibit a range of characteristics and serve to demonstrate the ability to differentiate levels of criticality for minerals with a variety of

[1]Investment decisions are based on a site-specific evaluation of ore reserves; capital and operating costs for alternative mining plan; the net present value (NPV) of after-tax cash flows, based on certain assumptions; and the sensitivity of NPV (or a related measure such as the rate of return) to changes in key parameters such as ore grade and tonnage, capital and operating costs, metal prices, and exchange rates.

properties and applications. Table 4.1 contains the criticality indicators for these three minerals (as well as eight additional candidates for criticality assessed in the next section).

A criticality matrix for a mineral can be constructed by first combining groups of applications and their percentage share of the total U.S. market for the mineral being evaluated (Table 4.2, columns 1 and 2 for example mineral X). The relative importance of the impact of a restriction in that mineral's supply for each application is then determined. The scale for this determination ranges between 1 and 4 (see also Figures 4.1 and 4.2), where 1 signifies the lowest impact for each application and 4 represents the highest impact for each application in the event of a restriction in mineral supply (Table 4.2, column 3). A composite "score" for the mineral for all applications is then determined using a simple weighting procedure, with the weighting factor based on the proportion (or percentage) of demand for that mineral. The composite score in column 4 for mineral X in the example of Table 4.2 would be $(0.27 \times 4) + (0.65 \times 4) + (0.08 \times 2) = 3.84$.

TABLE 4.2 Scoring the Vertical Axis of the Criticality Matrix for Example Mineral X

Application Group (End Uses) for Mineral X	Proportion of Total U.S. Market for Mineral X in Application	Impact of Supply Restriction (values of 1 to 4)	Weighted Score (product of columns 2 and 3)
Aerospace propulsion	0.27	4	1.08
Pigments	0.65	4	2.60
Biomedical devices	0.08	2	0.16
Overall importance in use	1.00[a]	NA	3.84[b]

NOTE: NA = not applicable.

[a]Total proportion will always equal 1.
[b]Final weighted score.

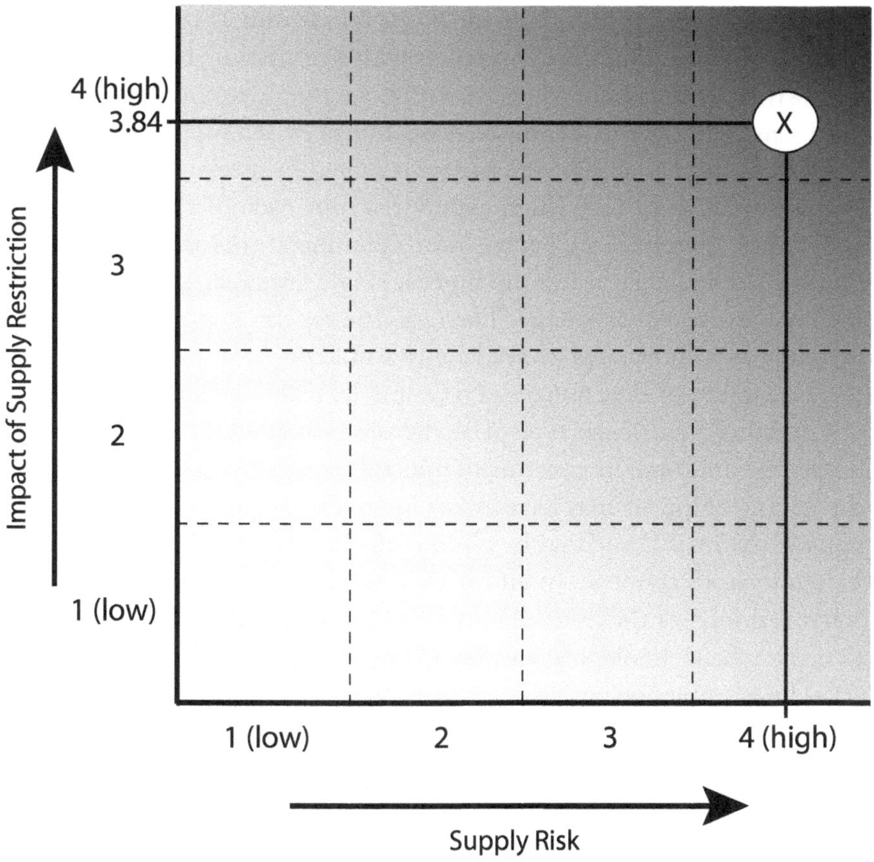

FIGURE 4.2 Matrix placement for example mineral X (see Table 4.1). The mineral's weighted composite score on the vertical axis of 3.84 (impact of supply disruption) coupled with a score of 4 on the availability axis (difficult to obtain due to technological impediments) place the mineral clearly in the critical range on the matrix.

This is essentially a weighted score between 1 and 4 for mineral X for all applications on the vertical axis of the matrix. As demonstrated below, the relative impact of a restriction in supply of a particular mineral has several dependencies. The relatively simple approach to the composite, weighted score for the vertical axis for a given mineral allows important end uses and end uses that command a high portion of the total use of that mineral to carry their weight into the evaluation. The scoring for the vertical axis considers not just impacts on products with a restriction, but also the ease of substitution for a mineral (in the event of a supply restriction) in a technical sense and the broader market consequences of a supply disruption.

The horizontal axis of the matrix is scored by assessing a value from 1 (low supply risk) to 4 (high supply risk) for each of the five areas of availability discussed in Chapter 3 for the mineral (not for each application sector) and then using the highest single availability factor score as the final horizontal axis score. The rationale for this approach is that any difficulty in a single area of availability will likely be a deciding factor in the availability of that mineral. For example, if the technological aspects of extracting a particular type of ore become so difficult (score of 4) as to hinder the ore's (and its constituent mineral's) extraction, then the fact that the mining of the ore may have low or no adverse environmental and social impacts (score of 1) and may be geologically known and abundant (score of 1) in numerous countries (political availability of 1) will not overcome the "unavailability" of the mineral in the short term due to the likely inability to bypass the technological barrier (Figure 4.2).

The examples below evaluate the criticality of individual minerals using data and assessments of the various factors influencing mineral use and availability. The treatments of each of the three detailed assessments are necessarily different in length because copper, as a single metal, is relatively more straightforward to discuss than groups of minerals such as the REs or PGMs. The committee was fairly certain that copper would not fall into the critical field before the analysis was conducted, but felt that an example with a high-volume, common metal such as copper could usefully illustrate application of the matrix and aid the reader in differentiating between the ideas that a mineral can be important, fundamental, or essential for many purposes, but may not be critical in the sense defined in this report. Other minerals likely to fall in this category include, but are not limited to, uranium (Box 4.1), bauxite (the mineral raw material for aluminum; Box 4.2), iron ore, and construction aggregate (Box 2.2) for which the committee did not undertake a detailed examination in this study.

BOX 4.1
Uranium Ore, Yellowcake, Enriched Uranium, and Nuclear Power

The abundance of uranium in the Earth's crust is similar to that of metals such as tin and tungsten, and many common rocks in different geological settings around the world, including granites and shales, contain fairly high uranium concentrations. Uranium ores are fairly easily removed from their host rocks and are treated by a variety of leaching solutions applied either through in situ wells or to mineralized rock that has been crushed and ground. In situ leaching does not impact deep aquifers because only uranium deposits submerged in shallow aquifers are mined in situ. The aquifer provides containment for the leaching solutions and is gradually drawn inward toward the extraction wells. After a well pattern is exhausted, aquifer water is extracted to flush the depleted deposit; the water is continuously treated and reinjected at impurity concentrations below regulatory (U.S. Nuclear Regulatory Commission) requirements.

A uranium compound, usually *yellowcake* (ammonium diuranate), is precipitated from the purified leach solution, dried, and shipped to a conversion plant where uranium hexafluoride is made. The hexafluoride, containing about 99.3 percent ^{238}U isotope and only 0.7 percent ^{235}U isotope, is then sent to an enrichment facility that generally uses gaseous diffusion to enrich ^{235}U to a fissionable concentration in the range of 3-6 percent.

The United States has 103 operating nuclear reactors in 64 power generating plants producing about 21 percent of its electric power. The operating cost of this power averages about 1.72 cents per kilowatt-hour compared to 2.21 cents for coal, 7.51 cents for natural gas, and 8.09 cents for oil. However, the capital cost for a nuclear plant can be as much as 1.5 times higher than a coal-fired plant and 2.5 times more than a natural gas-fired plant. The annual domestic refueling cycle requires about 70 million pounds of uranium, but U.S. production currently is only about 2 million pounds. Globally, the consumption of uranium is 170 million pounds, but 2005 production was only 108 million pounds, the balance deriving from a diminishing secondary supply of deactivated nuclear weapons. In reaction to this supply issue and to problems in overseas mines and mills, the price of yellowcake rose from $10 per pound in mid-2003 to $113 per pound in April 2007. These price increases have also spurred increased exploration activity for primary uranium sources around the world. Aided by advanced geophysical exploration tools, exploration for and discovery of new uranium deposits will likely maintain the sources that are otherwise being depleted through use.

SOURCE: Edwards and Oliver, 2000; Newton et al., 2006; World Nuclear Association, 2005.

BOX 4.2
Bauxite

Bauxite is an earthy mineral with approximate composition $Al_2O_3 \bullet 2H_2O$. Mined bauxite is treated at high temperature and pressure (Bayer Process) to produce an aluminum hydroxide precipitate that is dried and calcined, yielding highly purified *alumina*, Al_2O_3. Alumina is then reduced to molten aluminum metal (Al) by fused-salt electrolysis (Hall-Héroult Process).

Bauxite, which is also used for the manufacture of refractories and abrasives, was mined for alumina production in Arkansas, Alabama, and Georgia until the 1970s when reserves of high-grade material were exhausted. However, importation of bauxite and alumina had already begun and duties on both were abolished in 1971 by Public Law 92-151. Bayer plants continued to operate in the United States.

Since electrolytic reduction of alumina is very energy intensive, approximately 6 kilowatt-hour per pound of aluminum, the U.S. aluminum industry developed in the Pacific Northwest with the advent of cheap and abundant hydroelectric energy. That industry sector then became fully dependent on imported bauxite from Guinea, Jamaica, Guyana, and Brazil and imported alumina from Australia, Jamaica, and Suriname.

During the 1960s and 1970s, the domestic aluminum industry became concerned about potential economic strangulation by the "Bauxite Cartel." Every major producer invested tens of millions of dollars in the development of processes for production of alumina from non-bauxitic sources including *kaolin* clays from Georgia and *alunite* and *anorthosite* from the western United States. All production processes were successfully tested, but none was commercialized.

Copper

Vertical Ranking—Ease of Substitution and Impact of Supply Restriction on User Sectors

Copper was one of the first metals ever used by humans and continues to be extensively employed today. The uses of copper can be divided into five groups, as follows (see also Table 4.3; Joseph, 1999):

1. *Building and construction (about 55 percent of copper use):* electrical wire, plumbing and heating, air conditioning, commercial refrigeration, builder's hardware, and architectural uses;
2. *Generation and transmission of energy (about 15 percent of copper use):* electrical cable, step-up and step-down transformers;
3. *Transportation equipment (about 15 percent of copper use):* electrical distribution, alloys and cast components for road, rail, marine, air, and space vehicles;
4. *Machinery and equipment (about 10 percent of copper use):* computers, appliances, utensils, military ordnance, and industrial valves and fittings; and
5. *Telecommunications (about 5 percent of copper use):* communications cables and drop wiring; electronics.

In the building sector, the use of copper wiring is strongly increasing as energy use increases with new electronics and appliances and as buildings become larger on a per capita basis. Substitutes are not very satisfactory; thus copper appears to be a crucial component for the maintenance of modern building amenities. During the 1970s, attempts to substitute aluminum in residential wiring failed for reasons of reliability and safety.

TABLE 4.3 Relative Importance of End Use Applications for Copper

Application Group	Proportion of Total U.S. Market	Impact of Supply Restriction	Weighted Score
Building and construction	0.55	3	1.65
Energy provisioning	0.15	4	0.60
Transportation	0.15	2	0.30
Machinery and equipment	0.10	2	0.20
Telecommunications	0.05	1	0.05
Overall importance in use			2.80

NOTE: Proportion of total end use for each application was determined from Joseph, 1999.

Copper demand for hardware and architectural uses is stable, while that for plumbing is decreasing as plastics take an increasing market share. On the whole, building sector use of copper is increasing moderately, and for this reason, the committee evaluates the importance of an impact in supply restriction for this end use to be somewhat high (3) (see Table 4.3). In the energy sector, copper use is increasing strongly as a function of increased energy demand. Aluminum can capture some of the long-distance transmission use at decreased efficiency, but not the vital and copper-intense transformation use. Because of the great reliance at present on energy supply to the United States, we evaluate the importance of an impact in supply restriction for this end use to be high (4).

In the transport sector, copper use is flat to slightly increasing, as vehicles of all kinds become increasingly computer-controlled and as small motors for myriad uses are added. There are no good substitutes for copper in these uses, but good design can minimize the quantities required. Hybrid vehicles are an emerging use and increased demand for these vehicles would require an increase in the quantity of copper to satisfy current electrical and battery power design (see also Chapter 2). Although it is difficult to evaluate the total demand for hybrid vehicles in the future and, as a consequence, any increase in the demand for copper, the committee assesses the impact of a supply disruption for the transport sector as a whole to be somewhat low (2) at present. In the machinery and equipment sector, copper use is flat to slightly increasing. Excess use of material is being minimized, but this is balanced by increasing levels of computer control. Copper is important to these uses, but has potential substitutes; thus the committee suggests the impact to be somewhat low (2) for a disruption in the supply for this sector. In the telecommunications sector, copper use is flat to slightly decreasing. Around the world, more wireless telecommunications equipment is being installed. This is offset by the gradual replacement of copper cables with fiber optics. The criticality of copper in this sector is decreasing over time and is given a relatively low value (1) (Table 4.3).

Electricity drives every facet of today's society, and copper is the only reasonable metal for delivering electricity. As such, should a shortage occur, it would ripple quickly through the economy. The relatively high composite weighted score of 2.8 for copper reflects the magnitude of this effect.

HORIZONTAL RANKING—RISK TO COPPER SUPPLY OR COPPER AVAILABILITY

- *Geologic availability.* As discussed in Chapter 3, copper mining is concentrated in Chile, with Australia, Indonesia, Peru, and the United States also having significant operations. Many other countries mine copper at somewhat lower levels. This diversity of supply indicates that there are no immediate concerns for copper availability from a geologic perspective.
- *Technical availability.* Both open-pit and underground copper mining are as well developed for copper as for any nonrenewable resource. Recycling is generally straightforward because a large fraction of copper is used in pure form. Thus, no immediate concerns exist about copper availability from a technical perspective.
- *Environmental and social availability.* The regulatory environments in Chile and the other principal copper-producing countries are basically supportive of the industry, and copper mining and processing are less challenging environmentally than those for some other metals (e.g., gold, aluminum). This regulatory environment suggests that there is little significant concern for copper availability from an environmental perspective. The local communities in many of the principal copper-producing countries have not had a history of regarding mining activities as particularly problematic, although there is a moderate level of awareness of the operations. In the United States, some cases have been documented of differences between community interests and those of real estate, environmental, and mining interests (see also Chapter 3), and have led to mining being restricted in certain areas, and allowed in others. The use of recycled copper would be enhanced if personal electronics were recycled to a greater degree, which suggests the need for

more societal awareness worldwide. Overall, there are only modest concerns for copper availability from a social or environmental perspective.

- *Political availability.* With the possible exception of Indonesia, the principal copper-producing countries are politically stable. Particularly in view of the diverse set of countries capable of producing significant amounts of copper, there are no immediate concerns about copper availability from a geopolitical perspective.
- *Economic availability.* The diversity of copper producers dispels concerns of a few producers controlling prices, but the demand for copper has been strong in recent years as various large nations undergo industrial expansion. As a consequence, prices have risen accordingly.

On the basis of this information, the supply risk of copper is regarded as low to moderate and a value of (1) is suggested for all availability categories, yielding a value of (1) on the horizontal axis for each end use and for the composite use of copper. Combined with its weighted composite score of 2.8 on the vertical axis, Figure 4.3 shows that although copper is important to many aspects of daily activities and technologies, its availability does not make it a critical mineral at this time.

Rare Earths

Vertical Ranking—Ease of Substitution and Impact of Supply Restriction on User Sectors

The REs are subdivided into "light" and "heavier" elements reflecting their atomic numbers. The light REs, mainly cerium and lanthanum, comprise about 95 percent of the weight of all REs produced. The heavier REs command much higher prices and the commercially important ones include europium, neodymium, samarium, and gadolinium. Technically, all of the elements with atomic numbers 57 through 71 are REs, but some are not usually refined and occur simply as impurities in others. Although

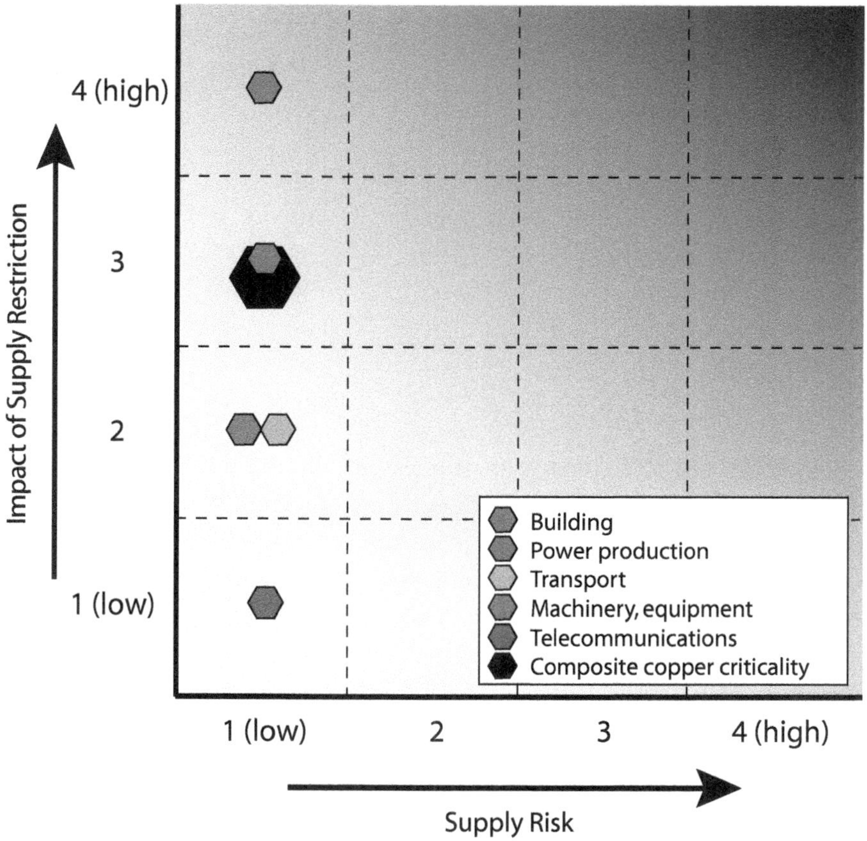

FIGURE 4.3 Criticality assessments for end use applications of copper and the composite criticality score for copper (black hexagon). Based on the committee's evaluation, copper is important, but not critical.

not a true RE according to its position in the periodic table of the elements, yttrium usually occurs with and is commonly included in the RE grouping.

End uses of REs and compounds in 2006 total a value in excess of $1 billion (Table 4.4; USGS, 2007).

TABLE 4.4 End Uses of Rare Earth Metals and Compounds

End Use	Percentage
Automotive catalytic converters	32
Metallurgical additives and alloys	21
Glass polishing and ceramics	14
Phosphors (television, monitors, radar, lighting)	10
Petroleum refining catalysts	8
Permanent magnets	2
Other	13

SOURCE: USGS, 2007.

In 1986 and 1996, end uses were essentially the same, but there has been a steady decline in use for catalytic fluid cracking of petroleum, while other applications have generally increased in volume and value. The diversity of end uses is primarily a reflection of the fact that the RE family includes many elements, some with only a few important applications. For the purposes of simplifying the evaluation of the impact of supply restriction of REs, the seven end uses for REs are here grouped into four categories having somewhat similar applications (Table 4.5).

TABLE 4.5 Relative Importance of End Use Applications for REs

Application Group	Proportion of Total U.S. Market	Impact of Supply Restriction	Weighted Score
Emission control, magnets, and electronics	0.44	4	1.76
Metallurgical, optical, and ceramics	0.35	3	1.05
Other	0.13	2	0.26
Petroleum refining	0.08	1	0.08
Composite, weighted score			3.15

NOTE: Proportion of total U.S. RE use for each application was determined from USGS (2007) information.

The chemical and physical properties and applications of REs make them sometimes important in the pure state as a metal or as the oxide, but usually one or more REs are used as alloys or compounds with non-RE elements (USGS, 2002). The following examples are offered in hopes of clarifying the remarkable characteristics of the RE family:

- Europium (Eu), as its oxide Eu_2O_3, serves as the red phosphor in color cathode ray tubes and liquid crystal displays and there is no substitute, despite prices on the order of $2000 per kilogram.
- Fiber-optic telecommunication cables incorporate periodically spaced lengths of erbium (Er)-doped fiber that function as laser amplifiers. Despite prices in the $1000 per kilogram range, there is no substitute because of erbium's unique optical properties.
- Cerium (Ce), as its oxide CeO_2, is one of the most abundant and cheapest REs, at a few dollars per kilogram, but it is used to polish virtually all mirrors and lenses because of its unique combination of physical and chemical attributes in an aqueous medium.
- One or more of the group of neodymium (Nd), samarium (Sa), gadolinium (Gd), dysprosium (Dy), and praseodymium (Pr), when alloyed with non-RE metals, can be used to make very-high-performance permanent magnets for miniaturized electronic and electrical devices with applications in automobiles, audio and video equipment, and military devices. There is no non-RE substitute in these cases or in multigigabyte portable disc drives ("memory sticks") or DVDs (digital versatile discs).

The growing high-tech applications were made possible by the extraordinary properties of combinations of REs (e.g., energy-efficient fluorescent lamps for industrial and institutional lighting that contain yttrium, lanthanum, cerium, europium, gadolinium, and/or terbium). Magnetic refrigeration is being developed as a potential replacement for gas-compression refrigeration and depends on the unusually high magnetic moments possessed by the trivalent ions of gadolinium, terbium (Tb), dysprosium, holmium (Ho), erbium, and thulium (Tm). If this technology is commercialized, it will enable reductions in energy consumption.

The use of cerium and other RE compounds grew in automotive catalytic converters in 2006. Demand also increased for mixed RE compounds and for RE metals and their alloys used in permanent magnets (USGS, 2007). Greater use of yttrium compounds in superconductors was noted as well. The USGS predicts that demand for REs will continue in many applications, particularly for automotive catalytic converters, permanent magnets, and rechargeable batteries (see also Chapter 2). In this first application category, substitutions may be possible for some electronic applications, but required performance parameters simply cannot be obtained with substitutes—for example, for neodymium or samarium in permanent magnets (see also Chapter 2). Given the increasing importance attached by our society to clean and healthful air, improved medical diagnostic tools, and electronic data transfer and communication, we suggest that the highest level of impact of supply restriction or importance in use (4) applies to the first application category, emission control, magnets, and electronics (see also Table 4.1).

The use of cerium compounds also grew in 2006 in the second application area for glass additives and polishing of optical glass. Higher consumption of yttrium compounds in 2006 was also noted for the manufacture of fiber optics, lasers, oxygen sensors, fluorescent lighting phosphors, color television screens, electronic thermometers, X-ray intensifying screens, and pigments. Demand increased for mixed RE compounds and for RE metals and their alloys used in base metal alloys, superalloys, pyrophoric alloys, lighter flints, and armaments. While insufficient information is available to make an accurate analysis of the ability of substitutes to perform satisfactorily in the second family of applications—metallurgical, optical, and ceramics—the committee suggests that growth in consumption of REs in these applications is powerful evidence that significantly improved performance has accrued from their adoption, yielding a ranking of (3).

A moderate degree of importance in use (2) reflects an admission that the committee is uncomfortable with either a high or a low assignment of criticality to an ill-defined application category ("other"). The use of RE chlorides in petroleum refining is decreasing, which is the primary rationale behind the assignment of a low value (1) to that use category.

The relatively high composite weighted score for REs of 3.15 (Table 4.5) reflects the diversity of applications for the RE family, the importance of those applications, and the steady growth in consumption and has led our committee to suggest that disruptions in the availability of REs would have a major negative impact on our quality of life. The relative importance ranking does not necessarily represent the average unit value of REs used in a particular application group. Also, the ranking does not reflect the potentially high intrinsic importance of the lowest-ranked applications. In our view, most of the applications are somewhat to very important since substitutes are generally less effective.

HORIZONTAL RANKING—RISK TO RE SUPPLY OR RE AVAILABILITY

- *Geologic availability.* The most important RE mineral families are *monazite* and *bastnäsite*, after Bastnäs, Sweden (also *bastnaesite*). Monazite, a by-product of beach sand, is a complex phosphate that can contain a variety of REs, along with thorium. Monazite is a by-product from mining carried out for the recovery of titanium and zirconium minerals. Minor amounts of monazite have been produced along the eastern seaboard of the United States. Bastnäsite is a complex fluorocarbonate that may contain either cerium and lanthanum or cerium and yttrium, along with an assortment of other REs. The less common and higher-unit-valued REs are by-products of cerium, lanthanum, yttrium, and titanium minerals (*ilmenite* and *rutile*) and/or *zircon*. The only recent domestic RE production has been from the cerium or lanthanum-type bastnäsite deposit at Mountain Pass, California. The Mountain Pass mine was last in operation in 2002 and has since been on a "care-and-maintenance" basis.
- *Technical availability.* The crustal concentrations of the REs are relatively high, rendering the term "rare earth" a practical misnomer; even the two least abundant, thulium and lutetium, are nearly 200 times as abundant as gold. However, in contrast to ordinary base and precious metals, the geochemical behavior of REs has

made concentration into exploitable mineral deposits a rare event. Discovering and developing a new, viable U.S. domestic RE production capability would be relatively difficult. Even if RE production at the Mountain Pass property resumes and continues, the product spectrum is narrow, chiefly comprising yttrium, cerium, europium oxide, a lanthanum-rich mixture of RE metals including praseodymium and neodymium, and a mixture of mainly samarium and gadolinium. The Mountain Pass carbonatite deposit is very depleted in RE elements heavier than dysprosium, compared with some Chinese deposits. Furthermore, the number of potentially viable RE deposits has been limited further by environmental and regulatory factors. For instance, monazite, which is the single most common RE-bearing mineral and one that occurs in the United States, contains elevated levels of thorium. Although thorium is only slightly radioactive, natural decay produces radium, which is highly radioactive, as are some subsequent decay daughters. The potential for accumulation of these products during processing poses significant challenges to the development of an economic mining and processing facility in the United States that could achieve fully permitted status. Largely due to the fact that most RE applications rely on alloying with other metals and because there are small quantities each of a broad spectrum of scrap types and compositions, recycling is not widely practiced, being confined primarily to magnet scrap, so secondary sources cannot easily be invoked.

- *Environmental and social availability.* Since 2002, some lanthanum concentrate has been produced from stockpiled ore at the Mountain Pass mine and has been sold to oil refineries as a cracking catalyst. In July 2004, a reclamation permit was approved, allowing some processing to take place, but RE separation facilities "are temporarily closed subject to the resolution of wastewater disposal issues" (Kohler, 2007, p. 73). According to the USGS (2002), "Even after the regulatory situation has been resolved, however, the long-term viability of Mountain Pass as a supplier of separated RE for high-technology applications is threatened by market factors." The USGS further points out that labor and regulatory costs are much lower in China than in the United States; the committee notes, however, that to some degree these costs differences will likely narrow as the Chinese economy develops and, in so doing, Chinese workers are able to command higher wages and salaries and as Chinese citizens demand stricter regulations for environmental quality and worker health and safety. Separation of REs at Mountain Pass is accomplished by a complex liquid ion exchange process with high operating costs; little is known about Chinese separation technology, but it may be simpler due to mineralogical and chemical differences between the plant feedstocks.

- *Political availability.* The United States is essentially 100 percent dependent on imported REs, with sources of refined metals and compounds as follows: China, 76 percent; France, 9 percent; Japan, 4 percent; Russia, 3 percent; and other, 8 percent. The predominance of U.S. importation of REs from a single country, coupled with that country's (in this case, China's) own growing demand for minerals (see also Chapter 2, Box 2.1), represents the potential for restrictions regarding future U.S. access to REs at any price.
- *Economic availability.* While the world reserve-to-annual production ratio is high at 715 (Table 4.1), it is deceptive because China accounted in 2006 for 97.6 percent of global mine production and the Chinese ratio of reserves to production is significantly lower, although still ample, at 225. Also, 21 percent of world reserves are in countries (United States and Australia) with no RE production and 25 percent are in countries accounting for only 0.3 percent of global mine production.

On the basis of the above considerations of the availability of REs, the committee considers supply risk for REs to be high for all applications, placing each of the above vertical rankings into the 4 category on the horizontal scale. These individual scores, combined with the vertical scores for each application, yield positions in the criticality matrix for each application area as shown in Figure 4.4. The composite, weighted criticality for REs is represented by the yellow circle and reflects the composite vertical score of 3.15 (Table 4.5) and the overall horizontal score for REs of 4. The final placement of REs on the diagram leads the committee to suggest that REs be considered critical minerals.

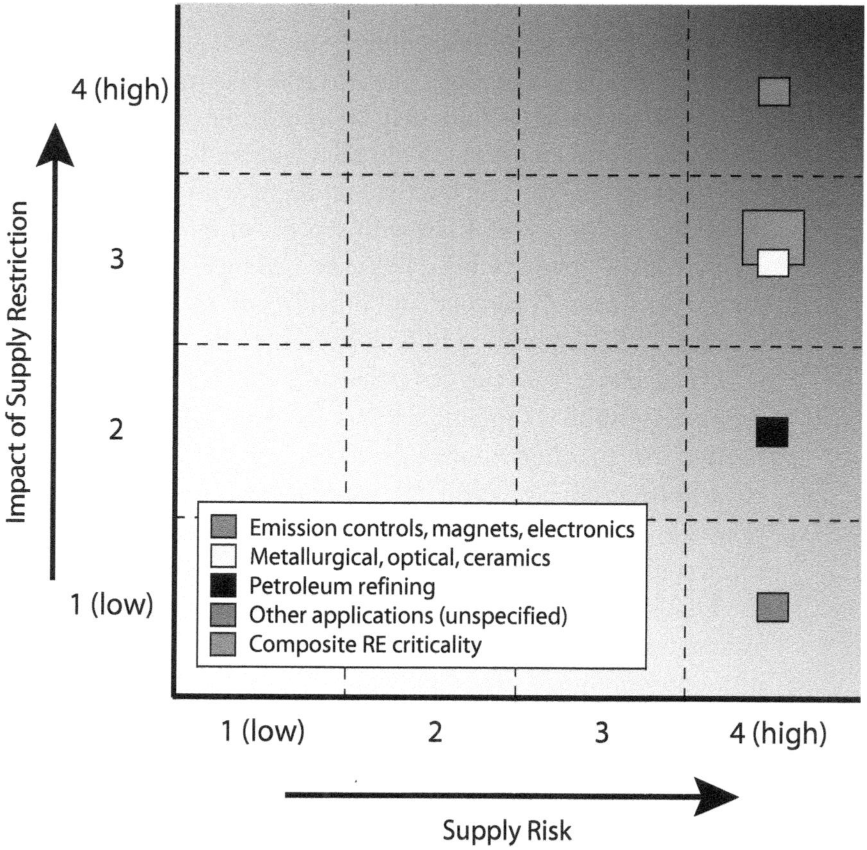

FIGURE 4.4 Criticality assessments for end use applications of REs and the composite criticality score for REs (blue square). Based on the committee's evaluation, REs are critical minerals.

Platinum Group Metals

Vertical Ranking—Ease of Substitution and Impact of Supply Restriction on User Sectors

The PGMs are platinum, palladium, rhodium, ruthenium, iridium, and osmium. The latter three see limited use and are not discussed here. PGMs and their alloys are characterized by high melting point, exceptional corrosion resistance, high strength, and ability to catalyze chemical reactions.

Current uses of PGMs can be classified in four main application groups: autocatalysts; industrial and other applications; fuel cells; and jewelry, dental, and electronics applications. Table 4.6 summarizes current PGM use (application) patterns, and these are described in more detail below.

Platinum and palladium are used in oxidative catalysts, aiding carbon monoxide and residual hydrocarbons to combine with oxygen to produce carbon dioxide and water vapor. Rhodium is used as a reducing catalyst to control nitrogen oxide emissions. In the automotive sector, catalyst composition depends on price, supply, fuel, and other factors. Palladium may be partially substituted for platinum in catalytic converters for gasoline vehicles. Reports indicate that some manufacturers had made progress in partially substituting palladium for platinum in catalytic converters for diesel vehicles, but the platinum:palladium ratio is now being increased in order to reduce particulate emissions (USGS, 2007). The committee was informed that palladium cannot be substituted for platinum in converters for diesel engines and there is no substitute for rhodium to control NO_x emissions (Herring, 2007). PGM use in this application has been reduced with the introduction of certain REs (USGS, 2007), but the supply of REs could also be vulnerable to disruptions (see previous section). An interruption in the supply of catalytic converters is a "no-build" condition since vehicles cannot be sold without a catalytic converter. Autocatalyst use, net of recycling, accounted for the majority of platinum, palladium, and rhodium used in 2006. This application group is expected to grow as governments and industry seek to improve air quality and fuel efficiency using lean-burn engine technologies that operate at higher temperature;

TABLE 4.6 Relative Importance of End Use Applications for PGMs

	Proportion of Total U.S. Market (2006)[a]			Relative Cost	Weighted Score		
Application Group	Platinum	Palladium	Rhodium	of Disruption	Platinum	Palladium	Rhodium
Autocatalysts for motor vehicle emission control	0.50	0.50	0.84	4	2.0	2.0	3.36
Industrial and other applications	0.25	0.05	0.16	4	1.0	0.20	0.64
Fuel cells for transportation or stationary applications	—	—	—	2	—	—	—
Jewelry, dental, and electronics applications	0.25	0.45	—	1	0.25	0.45	—
Overall score					3.25	2.65	4.00

[a]Johnson Matthey, 2007.

in the committee's judgment this application receives a high score (4) on the vertical axis.

In various industrial applications, PGMs are used to produce several high-volume industrial chemicals. Platinum and platinum-rhodium gauzes are used to catalyze partial oxidation of ammonia to nitric oxide, a raw material for fertilizers, explosives, and nitric acid. Platinum is also used in the production of sulfuric acid. The petrochemical industry uses platinum-supported catalysts in crude oil refining, reforming, and other processes to produce high-octane gasoline and aromatic compounds. Ruthenium dioxide is used to coat dimensionally stable titanium anodes used to produce chlorine, caustic soda, and sodium chlorate. For the industrial sector, PGM catalysts are essential determinants of product quality and yield for many high-volume chemicals. Supply disruptions could affect the availability and cost of materials, products, and fuels used by downstream industries, agribusinesses, and consumers. Other applications are being developed, for example, for electrowinning of copper. The committee suggests that the highest level of impact of supply restriction or importance in use (4) also applies to industrial applications for PGMs.

In fuel cells for transportation or stationary applications, platinum catalysts are used to help facilitate the production of electricity and water vapor by combining hydrogen and oxygen. Finely divided palladium is also used as a catalyst for hydrogenation and dehydrogenation reactions. At room temperature it can absorb up to 900 times its own volume of hydrogen, while hydrogen passes easily through heated palladium—a property that allows for purification and storage of hydrogen. Use data are not available for these emerging applications, which do not represent an important source of current demand. Because of the lack of data on these emerging applications, the committee was not comfortable in ranking the immediate impact of a supply restriction for these end uses higher than 2, but the committee acknowledges that demand could very well increase in the future. The committee also notes that supply restrictions could postpone the development and implementation of technologies that could reduce fossil fuel consumption, increase energy efficiency, reduce greenhouse gas emissions, and improve air quality.

Platinum is used in fine jewelry and is a constituent of white gold, a gold-platinum alloy. In jewelry, dental, and electronics applications, materials compete on the basis of absolute and relative prices, technical performance, aesthetics, exclusivity, and other factors. Platinum use in jewelry has decreased as a result of higher prices, while palladium use is growing as a result of lower prices. Substitutes are readily available for PGMs in jewelry and dental applications. Palladium alloys are used in dental applications, surgical instruments, watch springs, and jewelry. Palladium use in jewelry (as a substitute for white gold) is growing, particularly in China, and palladium is replacing rhodium-plated white gold in some luxury Swiss watches. The compound platinum dichloride is used in carbon monoxide detectors. Platinum, platinum alloys, and iridium are used as crucible materials to grow silicon, metal oxide, and other single crystals for semiconductor applications. A range of PGM alloys is used in low-voltage and low-energy contacts, thick- and thin-film circuits, thermocouples, furnace components, and electrodes. Less common PGMs are used to strengthen platinum, palladium, and other metals, including titanium. A low value (1) or low impact of supply restriction for PGMs is suggested for this application category.

U.S. PGM consumption in 2006 is estimated to have been approximately $1.83 billion (Table 4.1). PGMs are used in a range of essential applications, so a supply restriction would affect multiple industrial sectors that together represent an important proportion of U.S. economic activity. Barring a protracted supply restriction, autocatalyst and other existing uses for which substitution would be difficult or impossible that account for a substantial proportion of current global demand are expected to continue to grow. Emerging uses in hydrogen production and storage and fuel cell applications are also expected to grow. Because use patterns vary for each of the PGMs, calculating a composite, weighted score representative of all PGMs is not considered realistic. Calculating a composite score would also have to take into account the fact that data do not exist for some of the applications, although data are available for the more important applications (Table 4.6). Thus, the committee opted to derive a composite, weighted score for each of the three PGMs for which it has data. Based on the assigned scores and data for use in each sector, rhodium achieves a composite score of 4, platinum a score of 3.25, and palladium a composite score of 2.65, for the vertical axis.

Horizontal Ranking—Risk to PGM Supply or PGM Availability

- *Geologic availability.* The relative abundance of PGMs in the Earth's crust is orders of magnitude lower than that of base metals. Together with gold, rhenium, and tellurium, PGMs are among the rarest metals. PGMs occur in close association with one another and with nickel and copper, but economic concentrations are rare. From a geological perspective, concerns for PGM availability are assessed as relatively high (4).

  Platinum and palladium have the greatest economic importance of the PGMs and are found in the greatest quantities. Most primary production is from PGM ores in which the platinum content exceeds the palladium content and from which minor amounts of nickel, copper, and cobalt may be recovered. The other four PGMs are produced only as coproducts. Coproduct PGM recovery from nickel-copper-cobalt ores is a significant source, particularly in the Russian Federation and Canada.
- *Technical availability.* Technologies for primary extraction, processing and refining, and secondary recovery of PGMs are generally well established. Deep ore bodies can require advanced ground control and increased ventilation and air conditioning to provide a safe and hospitable environment for workers and equipment and ensure the physical integrity of the mine. Travel time for underground workers increases with depth, reducing labor productivity. The recycling technology for PGMs is among the most advanced of the metals, especially as far as industrial catalysts are concerned. Overall, there are no immediate concerns for PGM availability from a technical perspective (1), but technical factors are likely to increase capital and operating costs for new mines.
- *Environmental and social availability.* Regulatory environments in principal PGM-producing countries are generally supportive of

the industry. Local communities in the vicinity of mines, smelters, and refineries may bring social, environmental, or other impacts to the attention of operators and senior governments. Issues concerning the distribution of social and economic benefits or the mitigation of impacts may sometimes arise, but local communities are generally supportive of continued production and favor solutions that do not compromise current employment or future investment. Some deposits may not be developed as a result of resource or other conflicts, but most new mines are likely to be developed in established mining camps with a long history of PGM production, limited economic development alternatives, and a public consensus in favor of further mineral development. Labor availability is not a concern at this time, but the South African industry employs a significant number of workers from other African countries. Governments, industries, and communities throughout the region must confront a number of challenges, including HIV/AIDS awareness and treatment.

In the case of secondary supplies, PGM recycling from auto catalysts, electronics, and industrial catalysts is largely outside public view, and social factors are not involved.

From an environmental and social perspective, concerns for PGM availability are assessed as moderately low at this time (2).

- *Political availability.* Current PGM supply patterns (Table 4.7) show that, together, South Africa and the Russian Federation accounted for well over 80 percent of the 2006 global supply of platinum, palladium, and rhodium. Both countries have proven to be reliable suppliers.

PGMs are not transported by sea and are not vulnerable to maritime interdiction, but supplies could be restricted by local, regional, or other geopolitical developments. Inventories tend to be low, because of their high value and significant price and other risks. North American production would be inadequate to supply critical needs if the supply of platinum and rhodium from South Africa was interrupted. While the supply

TABLE 4.7 Current PGM Supply Patterns

	2006 Supply (%)		
Region	Platinum	Palladium	Rhodium
South Africa	78	34	89
Russian Federation	13	51	7
North America	5	11	2
Others	4	4	2

SOURCE: Johnson Matthey, 2007.

of palladium is somewhat more diversified, North American production would likely be inadequate to supply critical needs in case the supply of palladium from the Russian Federation or South Africa was interrupted. Furthermore, the only current U.S. producer of platinum and palladium, Stillwater Mining Company, is 55 percent owned by Norilsk Nickel, headquartered in Russia.

Significant refining, value-added processing, and recycling take place in Belgium, Norway, and the United Kingdom. Environmental concentrations have been attributed to autocatalyst applications, and these observations, together with the implementation of European Union legislation for the Registration, Evaluation and Authorization of Chemicals, are likely to trigger ecological and human health risk assessments that could lead to additional restrictions on emissions from production and recycling facilities. Use restrictions seem unlikely at this time due to the lack of known substitutes, but regulatory costs are likely to be passed on to users.

A high degree of import dependence for PGMs or other minerals and metals is not, in itself, cause for concern (see also Table 4.1). Increased trade and investment flows contribute to economic growth and prosperity, for both the United States and its trading partners. In the particular case of PGMs, the United States is highly dependent on imports that originate largely in the Russian Federation and South Africa. As outlined earlier in this chapter, U.S. imports support investment in those countries and generate social and economic benefits there, but they expose a range of U.S. industries to political, economic, and other risks that vary according to the particular situation.

From a political perspective, concerns about PGM availability are assessed as relatively high at this time (4).

- *Economic availability.* About 10 mining companies and a smaller number of smelting and refining companies account for substantially all primary PGM production. The same smelters and refineries play an important role in the recovery of PGMs from secondary sources. PGM coproducts are an important source of revenue for certain nickel-copper-cobalt producers in the Russian Federation and Canada. Production of PGMs from those sources is largely nondiscretionary and is driven primarily by decisions related to the main products. The largest nickel producers in both countries have sufficient reserves to support continued operation for decades. Although production of PGMs is unlikely to be reduced in response to lower prices, it is equally true that higher prices are unlikely to stimulate increased production from those sources. Although coproduction is a factor for all PGMs, base metal production accounts for a higher proportion of palladium than platinum production. Since the supply is relatively insensitive to price, the result can be greater price volatility for palladium. A growing proportion of nickel production will originate from lateritic deposits (containing little or no PGMs) in the future, so new nickel mine developments will not necessarily increase PGM production.

  As noted above, technical factors are tending to increase capital and operating costs for new mines. Producers will take a disciplined approach to new capital investment to achieve targeted rates of return despite significant price, foreign exchange, and other risks. Finally, the largest producers operate in jurisdictions with legal requirements that may not promote market transparency and disclosure or discourage anticompetitive or monopolistic behavior.

> Production of PGM alloys and value-added PGM products is controlled by a very small number of companies that may be in a position to exercise considerable market power, although to a large extent those activities are conducted in countries with legal requirements in place to discourage anticompetitive behavior.
>
> For all of these reasons, from an economic perspective concerns for PGM availability are assessed as relatively high (4).

The world reserve-to-production ratio and the world reserve base-to-production ratio integrate certain aspects of geological, technical, and economic availability. The ratios therefore provide an indication of the long-term availability of PGMs from primary sources. It is difficult to anticipate future exploratory success, prices, costs, exchange rates, or production levels, all of which affect the resulting ratios. The underlying database may provide insight into the economic outlook for existing mines, but the ratios provide no insight into market dynamics.

As Table 4.1 shows, platinum is produced primarily by primary platinum mines with by-products that may include other PGMs, nickel, copper, and cobalt. Significant platinum production is a by-product of nickel-copper-cobalt and palladium production. In contrast, palladium is produced primarily as a by-product of platinum and nickel-copper-cobalt production, with limited production from primary palladium mines; other PGMs are produced exclusively as by-products. Short-term platinum supply and price are driven primarily by fundamental factors such as cash operating costs, exchange rates, demand, and inventory. By-product dependence makes the supply of palladium and other PGMs less price sensitive and the prices more volatile, as production levels respond primarily to prices of the main product. If demand for palladium and other PGMs was to grow more rapidly than demand for platinum, this could lead to higher prices for palladium and other PGMs. Since demand for nickel is increasingly supplied from lateritic deposits, the supply of palladium and other PGMs should gradually become more closely linked to primary platinum production in South Africa and to recycling activity.

U.S. secondary production from old scrap, as a percentage of U.S. apparent consumption, is assessed as significant for PGMs. The high value of PGMs, the controls on disposal of solid waste that exhibits hazard characteristics, and the concerns about potential liability are important drivers for PGM recycling and for collection and regeneration of spent catalysts. Available data indicate that a higher proportion of current autocatalyst demand for platinum and palladium is satisfied by recovered PGMs in North America than in Europe, Japan, or the rest of the world (Johnson Matthey, 2007). The development of an infrastructure to collect and recover PGMs in North America has led to the development of a new economic activity that reduces import dependence and the associated risks. While this may suggest that opportunities for improvement are limited in North America, an opportunity could exist to further increase collection and recovery and to lever this strength by working with other countries to establish an efficient collection infrastructure, importing PGM scrap for processing in the United States or another country and developing offshore recycling capacity as the supply of scrap grows. While such an initiative could best be undertaken by companies, government action may be needed in the case of existing international agreements between nations. Because certain materials may be subject to the Basel Convention, for example, parties to the convention may be unable to ship scrap to a U.S. facility for processing because the United States has not ratified the convention; an exception could occur under the terms of an agreement or arrangement that conforms to Article 11 to the convention.[2] Alternately, a business could site a facility in Asia or in another country that is a party to the Basel Convention. Although secondary production from old scrap leaves the U.S. economy less exposed to political, economic, and other risks than it might otherwise be, and opportunities may exist to increase this activity, the United States remains vunerable to the risk of a restriction in the supply of PGMs.

[2]In practical terms this could require a bilateral agreement, unless the exporting state is also a member of the Organisation for Economic Co-operation and Development.

On the basis of these assessments of geologic, technical, environmental and social, political, and economic availability, the probability of a restriction in the supply of PGMs is assessed as relatively high (4). Based on the impacts of a potential PGM supply restriction, which are assessed as relatively high (4), and the relatively high probability of a PGM supply restriction, PGMs are considered to be critical minerals (Figure 4.5).

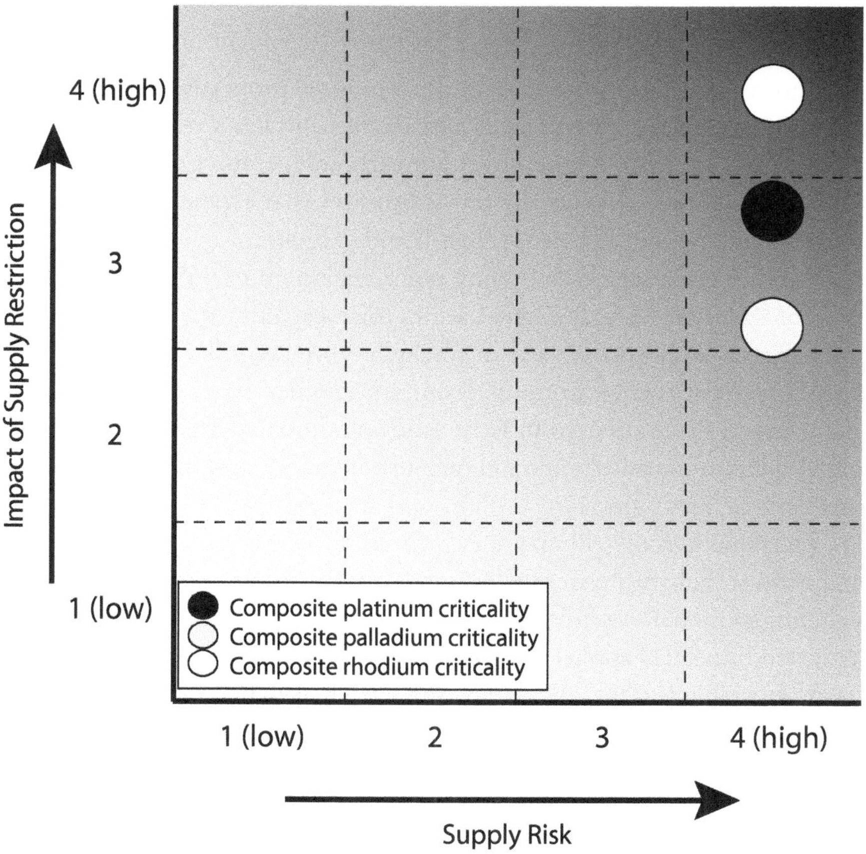

FIGURE 4.5 Criticality assessments for PGMs platinum, palladium, and rhodium.

OTHER CRITICAL MINERAL CANDIDATES

In addition to the minerals discussed above, a number of other minerals have been suggested as potentially critical in the recent past. In less complete fashion than has been done above, the committee presents a discussion of the criticality aspects of eight of these minerals.

Gallium

Vertical Ranking—Ease of Substitution and Impact of Supply Restriction on User Sectors

Gallium is a soft, silvery metal with many unusual properties. It is stable in air and water but reacts with acids and alkalies and has a very low melting point. Gallium is also a liquid over a greater temperature range than any other element and is one of the few substances that expands as it freezes, making solid gallium less dense than liquid gallium.

Gallium has a very high boiling point, making it useful for high-temperature thermometers. It is also used in mirror making and alloying with other metals. Gallium is used in research—for example, solar neutrino detection experiments. Some gallium compounds have semiconductor characteristics and are used in light-emitting diodes (LEDs) and transistors. A gallium-arsenic compound (gallium arsenide) can convert electricity directly into laser light.

The main uses of gallium are in high-technology, defense, and medical applications. Substitutes for gallium tend to be confined and very specific, including indium phosphide components in laser applications; some organic compounds in LED applications; and silicon in solar cell applications. Domestic recycling is not widely practiced, with the main source of recycling resulting from the reprocessing of new scrap generated in the manufacture of gallium arsenide devices. End use applications, their weighted scores, and the composite score for gallium are presented in Table 4.8.

TABLE 4.8 Relative Importance of End Use Applications for Gallium

Application Group	Proportion of Total U.S. Market (2006)[a]	Impact of Supply Restriction	Weighted Score
Integrated circuits (technology)	0.63	3	1.89
Optoelectronic devices (cell phones, backlights, flashes)	0.22	2	0.44
Research, specialty alloys, other	0.15	1	0.15
Overall importance in use			2.48

[a]USGS, 2007.

Horizontal Ranking—Risk to Gallium Supply

Gallium rarely occurs in minerals except in trace amounts. Bauxite and the mineral *sphalerite*, as well as coal, often contain gallium as an impurity. No gallium is mined; it is obtained as a by-product of mining and processing other metals—notably aluminum, zinc, and copper—and is produced in any nation that produces these metals. In 2006, the United States was 99 percent reliant on foreign sources for gallium. Import sources include China, 37 percent; Japan, 17 percent; Ukraine, 12 percent; Russia, 10 percent; and other, 24 percent (USGS, 2007). In 2000, foreign reliance was quantified as "some." In 2004, it was 99 percent, maintaining that level into 2006.

We consider supply risk for gallium to be elevated to high (3), based on supply sources, substitutability, and recycling constraints. The composite criticality for gallium is located together with the other seven "criticality candidates" on Figure 4.6.

Indium

Vertical Ranking—Ease of Substitution and Impact of Supply Restriction on User Sectors

Indium is a very soft, stable metal, being unaffected by water and air, but it does react with most acids. Its major use has come to be as a coating material for flat-panel displays, for which no adequate substitutes are currently available. The very rapid increase in the production of these displays has led to some supply shortages.

Indium is also used to make mirrors, often replacing silver because of its noncorroding characteristics. It has a low melting point and is used in making low melting alloys for safety devices and solders. Indium, like gallium, remains a liquid over a large temperature range. Some indium compounds are used in transistors, photoconductors, photocells, and thermistors.

Indium has substitutes for most of its applications, although the substitutes lead to loss in efficiencies and/or product characteristics. Silicon has largely replaced indium (and germanium) in transistors. Gallium (more expensive) can replace indium in some alloys; silver-zinc oxides or titanium oxides can substitute for indium in glass coatings. Indium phosphide can be replaced by gallium arsenide in solar cells and in many semiconductor applications. Gold-tin is the only substitute for indium-tin solders used in printed circuitry (e.g., for tunable lasers). Most of the substitutions occur with other commodities that are challenged by their own supply source constraints (gallium, for example). The amount of indium recycling in the United States is very small due to the lack of infrastructure to collect indium-bearing products. End use applications, their weighted scores, and the composite score for indium are presented in Table 4.9.

Horizontal Ranking—Risk to Indium Supply

Some indium is found in pure form, and it occurs in only a few minerals, such as *indite*. However, almost all indium is obtained as a by-product of zinc processing and is recovered from residues left from electrolytic refin-

TABLE 4.9 Relative Importance of End Use Applications for Indium

Application Group	Proportion of Total U.S. Market (2006)[a]	Impact of Supply Disruption	Weighted Score
Coatings	0.70	4	2.80
Electrical components; semiconductors	0.12	3	0.36
Solders and alloys	0.12	2	0.24
Research; other	0.06	1	0.06
Overall importance in use			3.46

[a]USGS, 2007.

ing of zinc. Worldwide distribution of indium-bearing ores and deposits includes Asia (China and Russia, primarily), Europe, North and South America, Australia, and South Africa (Stevens, 2007). Because zinc, copper, lead, and tin ores, among others, are the common ores that also host indium, the technology for extraction of indium is dependent on development in these mining sectors, and indium availability is not currently considered a problem. As a by-product of zinc (or other) ores, the technical challenges, particularly with higher prices for indium, arise in making the refining process as efficient as possible. By-product extraction, enhanced recovery from this extraction, purification, and refinement before the indium is incorporated in its end applications are all technologically established processes that are undergoing continued improvements to increase the efficiency of the recovery of indium from the virgin ore, tailings, and recycling of end products (Stevens, 2007). In 2006, the United States was 100 percent reliant on foreign sources for indium. Import sources include: China, 44 percent; Canada, 22 percent; Japan, 15 percent; Russia, 5 percent; and other, 14 percent (USGS, 2007). In 2000, there was considered to be some foreign reliance; in 2001, it was 95 percent; and by 2006, the United States was 100 percent reliant on foreign sources. The risk of supply restriction through export quotas from one or another supplier to the

United States is important to consider in evaluation of indium's continued availability.

Indium production rose from 60 metric tons annually in 1970 to 400 metric tons in 2005 in response to increased demand, although industry production was not always able to anticipate the rate of increase in demand and sometimes lagged behind (Stevens, 2007). In partial response to the increase in demand, between 2003 and 2006, the price of indium rose from about $100 per kilogram to $980 per kilogram, attributed largely to the great increase in demand for indium tin oxide—a compound used for liquid crystal displays (Stevens, 2007; USGS, 2007). Although Stevens (2007) suggested that indium would be both available and economical in the long term, the committee considers the short-term volatility to add considerably to the supply risk for indium. The short-term supply risk for indium is therefore considered to be elevated to high (3), based on supply sources, substitutability, and recycling constraints. The composite criticality for indium is located together with the other seven criticality candidates on Figure 4.6.

Lithium

Vertical Ranking—Ease of Substitution and Impact of Supply Restriction on User Sectors

Lithium is important in the manufacture of glass, ceramics, and aluminum. It is also used in making synthetic rubber, greases, and other lubricants. In these applications, potassium can be substituted for lithium. Nonrechargeable lithium batteries are used in products such as calculators, cameras, and computers. The largest potential for growth is in the manufacture of batteries, especially rechargeable batteries used in video cameras, portable computers, and telephones.

It is in future applications that supplies and costs of lithium will be more important. Lithium is a key to the development of new battery technology. Lithium ion batteries have an exceptionally good energy-to-weight ratio and are therefore valuable in reducing overall automobile weight and thereby improving fuel efficiency. Lithium ion batteries are not used in current hybrid car models, but this technology, available at a reasonable cost, may be particularly important if hybrid vehicles become an important factor in our transportation system. End use applications, their weighted scores, and the composite score for lithium are presented in Table 4.10. At present we conclude that the current importance of lithium would be low, but believe that this element must be followed closely as this situation could change in the future.

Horizontal Ranking—Risk to Lithium Supply

Lithium is widely distributed in Earth's crust, but at low concentrations. For economic reasons it is produced primarily from subsurface brines, rather than being mined from hard rock. The United States is the leading consumer of lithium-based products. Only one company in the United States produces lithium compounds. The leading producers of lithium ore are Chile and Argentina. A preferred source of lithium is recovery from brine. However, the United States has only one brine recovery operation in Nevada.

TABLE 4.10 Relative Importance of End Use Applications for Lithium

Application Group	Proportion of Total U.S. Market[a]	Impact of Supply Disruption	Weighted Score
Ceramics and glass	0.21	2	0.42
Batteries	0.19	3	0.57
Lubricants	0.16	1	0.16
Pharmaceuticals	0.09	2	0.18
Other	0.27	2	0.54
Overall importance			1.87

[a]USGS, 2007.

The committee considers supply risk for lithium to be at least moderate (2), based on supply sources (dominance by two large companies), substitutability, and recycling constraints. The composite criticality for lithium is located together with the other seven criticality candidates on Figure 4.6.

Manganese

Vertical Ranking—Ease of Substitution and Impact of Supply Restriction on User Sectors

Manganese is a silver-white metal that can be hardened by alloying with carbon. Manganese generally is used to harden steel and other iron alloys, and it has important minor uses such as hardening aluminum for the fabrication of stiff lids for otherwise very ductile aluminum beverage cans. Electrolytic manganese dioxide is the depolarizer in dry cell batteries, and continuous quality improvements during the last 30 years have increased battery shelf life from 1-2 years to 7-10 years.

In the end uses listed in Table 4.11, there are no satisfactory substitutes for manganese. It has been considered a strategic metal for the U.S. military since the 1930s, and newer applications add a degree of criticality. The committee has therefore concluded that the impact of supply restriction and weighted rankings are as shown in Table 4.11, with dry cell batteries perhaps being somewhat more important in use than nonsteel alloys, and fertilizers and micronutrients being slightly less important in use for this mineral.

Horizontal Ranking—Risk to Manganese Supply

Manganese is relatively abundant in the Earth's crust and also exists on some areas of the Pacific Ocean floor, as well to a lesser degree in the Indian and Atlantic Oceans. The most common ore minerals are the manganese oxides (e.g., *pyrolusite*) and the carbonate *rhodochrosite*. The likelihood of discovering high-grade domestic ore reserves is negligible. Manganese

TABLE 4.11 Relative Importance of End Use Applications for Manganese

Application Group	Proportion of Total U.S. Market[a]	Impact of Supply Restriction	Weighted Score
Steel and cast iron	0.90	4	3.6
Other alloys	0.05	3	0.15
Dry cell batteries	0.03	3	0.09
Fertilizers and micronutrients	0.02	3	0.06
Overall importance in use			3.90

[a]USGS, 2007.

deposits in the United States are now too low grade for commercial exploitation, although the Butte and Philipsburg Districts in Montana were significant producers of ferromanganese and battery-grade manganese until the 1960s when minable ores were exhausted. Since then, essentially all manganese has been imported in the form of manganese ore, ferromanganese, and silicomanganese. Recycling is minor, although some manganese is recovered from steelmaking slag. The United States is essentially 100 percent dependent on imported manganese, with 72 percent of all manganese ore being imported from Gabon and 67 percent of the ferromanganese coming from South Africa, China, and Brazil. Reported U.S. consumption is closely tied to steel consumption and has generally increased during the last 50 years, totaling 870,000 metric tons in 2006. The average price for 46-48 percent manganese ore in 2006 was $3.61 per metric ton unit (10 kilograms), or about $0.164 per pound (USGS, 1997, 2007), yielding a consumed value of $314 million.

Based on these considerations, we consider supply risk for manganese to be fairly high for all applications. The composite criticality for manganese is located together with the other seven "criticality candidates" on Figure 4.6.

Niobium

Vertical Ranking—Ease of Substitution and Impact of Supply Restriction on User Sectors

Niobium[3] is silvery in appearance with a high melting point of 2468°C. Applications in steelmaking include carbon steels; high-strength, low-alloy (HSLA) steels; and stainless and heat-resistant steels. The aerospace industry is a major user of superalloys and niobium metal (Table 4.12). Since 1995 the percentage of niobium consumption in HSLA steels has dropped by two-thirds and the percentage consumed in superalloys has nearly doubled from 18 percent.

Niobium's uses are specialized and important, and substitution generally carries with it reduced performance or higher cost. Nonetheless, molybdenum and vanadium in HSLA steels; tantalum and titanium in stainless steels; and ceramics, molybdenum, tantalum, and tungsten in high-temperature applications are potential substitutes. Chemical uses of niobium are not very important and not of major concern here. The committee therefore believes that the appropriate criticalities and weighted rankings are those shown in Table 4.12 with steels perhaps being slightly more important in use than chemicals.

Horizontal Ranking—Risk to Niobium Supply

Niobium is an uncommon element, usually found with tantalum and occurring as the iron niobium oxide, *columbite*, and the niobium oxides, *pyrochlore* and *samarskite*, with the largest deposits in Brazil and Canada. No significant U.S. mine production has been reported since 1959 (South Dakota) and the likelihood of discovering domestic ore reserves is negligible. Recycling data are not available, but the USGS (2007) has estimated

[3]U.S. industry often uses a second name for niobium (Nb)—columbium, as identified by Hatchett in 1801. However, the International Union of Pure and Applied Chemistry decided in 1950 that niobium is the accepted name for element 41. The committee has opted to retain this internationally accepted name for purposes of this report.

TABLE 4.12 Relative Importance of End Use Applications for Niobium

Application Group	Proportion of Total U.S. Market[a]	Impact of Supply Restriction	Weighted Score
Steels	0.67	3	2.01
Superalloys and niobium metal	0.32	4	1.28
Chemicals and other	0.01	3	0.03
Overall importance in use			3.32

[a]USGS, 2007.

that recycling may account for as much as 20 percent of apparent consumption. The United States is essentially 100 percent dependent on imported niobium, with 80 percent being imported from Brazil and 10 percent from Canada. Trade agreements and established relationships with these nations do not indicate any significant risk to supply at this time. Reported U.S. consumption has tripled during the last 15 years, totaling 10,300 metric tons in 2006. The average price for ferroniobium in 2006 was $7.62 per pound of contained niobium (USGS, 1997, 2007), resulting in a consumed value of $173 million.

With these factors taken into consideration, the committee considers the supply risk for niobium to be fairly high for all applications (3). The composite criticality for niobium is located together with the other seven criticality candidates on Figure 4.6.

Tantalum

Vertical Ranking—Ease of Substitution and Impact of Supply Restriction on User Sectors

Tantalum is a metallic element having high ductility, is a good conductor of electricity, and is highly resistant to corrosion by acids. Its major use is in the production of electronic components, mainly capacitors. As noted in Chapter 1, a tantalum-containing compound is a significant component in the manufacture of capacitors for the base stations that are part of the cellular telephone network, operating at 2.2 gigahertz, as well as almost all other modern electronics (Table 4.13).

Although considerable research has been carried out, to date no effective substitute has been found for tantalum in this application. Tantalum can be substituted for, but only with a loss in performance (e.g., more dropped calls, shorter battery life, reduced electronics performance). Domestic recycling of tantalum is essentially nonexistent, because of its use in very small quantities in a large number of products. The committee therefore determines that the appropriate criticalities and weighted rankings are those shown in Table 4.13.

Horizontal Ranking—Risk to Tantalum Supply

Tantalum is not mined in the United States; most of our imports come from Australia. Other major sources of tantalum are Canada and Brazil. Because most of the tantalum mined in the world is used in capacitors, compared to the relatively small quantities needed for this market, cell phone manufacturers can be held captive to high prices. As an example, the price of tantalum has increased 1000 percent since 1955. Australia, Canada, and Brazil are considered reliable suppliers. In terms of the availability of tantalum, it is not so much a question of obtaining the material, but its price. The committee would thus rank supply risk as moderate (2).

TABLE 4.13 Relative Importance of End-Use Applications for Tantalum

Application Group	Proportion of Total U.S. Market[a]	Impact of Supply Restriction	Weighted Score
Capacitors	0.65	3	1.95
Specialty alloys, other	0.35	2	0.70
Overall importance in use			2.65

[a]USGS, 2007.

The composite criticality for tantalum is located together with the other seven criticality candidates on Figure 4.6.

Titanium

Vertical Ranking—Ease of Substitution and Impact of Supply Restriction on User Sectors

The element titanium has important uses in two distinct forms: titanium dioxide (TiO_2) and titanium metal. Titanium dioxide is a vital component in paints and pigments—its overwhelming use on a mass basis. Titanium metal is quite resistant to corrosion and has a high melting temperature. While its strength is similar to steel, it is 45 percent lighter, and titanium alloys can be twice as strong as aluminum alloys. As a strong, lightweight metal, titanium is thus an important component in aerospace applications, which have very few substitutes. In 2004, an estimated 60 percent of the titanium metal (as opposed to the oxide for pigment) used in the United States went into aerospace applications. The remaining 40 percent was used in such applications as armor, chemical processing, marine, medical, and power generation, for example. Because it is compatible with the human body, titanium is often used in surgical instruments and medical implants.

Titanium presents an interesting example of the distinction between total uses and the impact on different user sectors (Table 4.14). In its dominant use, pigments, reasonable substitutes are available; as a result, the overall importance according to the algorithm used in this report is quite moderate. However, the metal is absolutely crucial to the manufacture of aircraft and other high-technology products, and these products are crucial to the U.S. economy and to its balance of payments. Restrictions on supply would thus have major implications for one sector that is (relatively) not a big titanium user, but is very important to the U.S. economy. Conversely, supply restrictions would be much easier for the dominant pigments sector to address.

TABLE 4.14 Relative Importance of End-Use Applications for Titanium

Application Group	Proportion of Total U.S. Market[a]	Impact of Supply Disruption	Weighted Score
Pigments	0.95	2	1.90
Aerospace, high technology	0.05	4	0.20
Total			2.10

[a]USGS, 2007.

Horizontal Ranking—Risk to Titanium Supply

Titanium is the ninth most common element in the Earth's crust. Titanium occurs in the minerals *rutile* and *ilmenite*; ilmenite provides about 90 percent of the titanium every year. Chlorinating rutile and reducing the product to titanium sponge using magnesium metal produces titanium metal. The sponge is converted to titanium metal in an ingot form in an electric arc furnace. In 2006, the United States was 67 percent reliant on foreign sources for titanium. The major import sources were Australia and Canada (USGS, 2007).

The committee considers supply risk for titanium to be moderate (2), based on supply sources, substitutability, and recycling constraints. The composite criticality for titanium is located together with the other seven criticality candidates on Figure 4.6.

Vanadium

Vertical Ranking—Ease of Substitution and Impact of Supply Restriction on User Sectors

Vanadium is a very hard metal with a relatively high melting point of 1895°C. Addition of vanadium during steelmaking results in the formation of a finely dispersed vanadium carbide phase that is very hard and wear resistant. High-performance titanium alloys for aerospace applications typically contain 4 percent vanadium and 6 percent aluminum (Table 4.15).

TABLE 4.15 End Uses of Vanadium and Its Compounds[a]

End Uses	Percentage
Carbon steels	25
High-alloy steels	27
High-strength, low-alloy steels	27
Other alloys (e.g., with titanium)	11
Catalysts and chemicals	10

[a]Primarily ferrovanadium and vanadium pentoxide.

SOURCE: USGS, 2007.

Niobium, manganese, molybdenum, titanium, and tungsten are to some degree interchangeable with vanadium as alloying elements in steel. In some catalytic applications, platinum and nickel can be substituted for vanadium, but usually at a higher cost. In aerospace titanium alloys, there is currently no acceptable substitute for vanadium. These considerations lead the committee to conclude that the impacts of supply disruption and weighted rankings are as shown in Table 4.16, where the alloys have been combined into a single end use category.

Horizontal Ranking—Risk to Vanadium Supply

Vanadium is found in a broad spectrum of minerals distributed in many countries, as well as in Canadian tar sands and crude oil produced by Mexico and Venezuela. Vanadium has been produced domestically for decades, primarily from wastes and residues such as slags and boiler ash, but domestic production is dependent strongly on prices. In 1996, nine vanadium producers had eight active extraction operations whose feed comprised ferrophosphorus slag in Idaho, vanadium-bearing iron slag, petroleum residues, spent catalysts, and boiler ash, for example, from ships burning Mexican fuel oil. In 2006, eight producers of vanadium-bearing materials existed, but all used imported feedstock and semirefined compounds.

TABLE 4.16 Relative Importance of End Use Applications for Vanadium

Application Group	Proportion of Total U.S. Market[a]	Impact of Supply Restriction	Weighted Score
Steel alloys	0.79	2	1.58
Other alloys	0.11	4	0.44
Catalysts and chemicals	0.10	3	0.30
Overall importance in use			2.32

[a]USGS, 2007.

Recycling is not widely practiced, although the extent of recycling would probably increase if warranted by restricted supply or sustained high prices. However, ample domestic sources of vanadium exist and processing plants could probably be restarted if economically justified.

Several mothballed U.S. uranium mills are likely to be reactivated within the next 3 years and at least one could have a vanadium by-product. One mothballed conventional uranium mill is located in Utah and the other in Wyoming. One operating plant is in Utah and one is in Colorado. A mothballed plant in Idaho that produced vanadium metal and vanadium pentoxide from ferrophosphorus slag may be capable of reactivation. Uranium can also be recovered as a by-product of phosphoric acid production, and there are two plants each in Florida and Louisiana on standby status.

The United States at present is essentially 100 percent dependent on imported vanadium feedstocks, with 74 percent of ferrovanadium from the Czech Republic and 82 percent of vanadium pentoxide from South Africa. In mid-2006, Anglo American was reportedly in the process of selling its 79 percent controlling interest in Highveld Steel and Vanadium Corporation Limited to Russia's Evraz group for U.S. $678 million (Moore, 2006). The committee has not confirmed this transaction, but it could potentially increase supply risk. Reported U.S. consumption for the last 15 years has been fairly constant in the range 3000-4300 metric tons annually. From 1991 through 2003, the price of vanadium pentoxide was in the range

$1.34-2.95 per pound, but prices then rose sharply to a high of $16.28 in 2005, falling to a 2006 average of $8.08 (USGS, 1997, 2007). Thus, total 2006 value was about $68 million. The committee considers supply risk for vanadium to be only moderate (2) for all applications, and the composite criticality for vanadium is located together with the other seven criticality candidates on Figure 4.6.

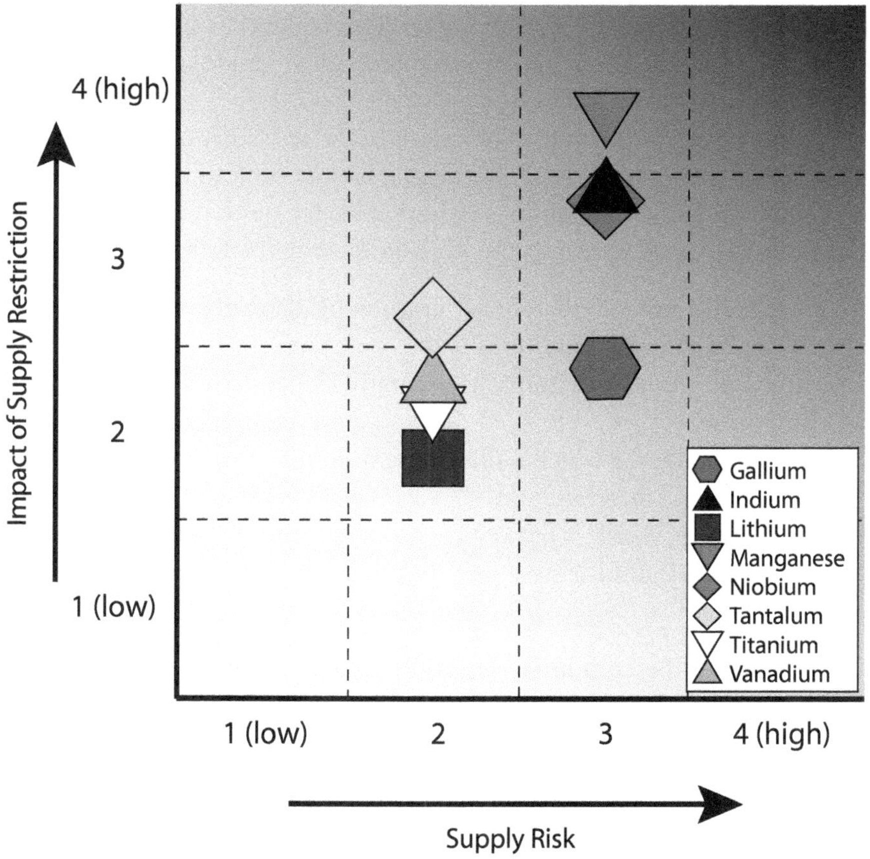

FIGURE 4.6 Criticality matrix for the eight candidate minerals discussed above. Of the eight, niobium, indium, manganese, and potentially gallium stand out as minerals of potential concern to the U.S. economy.

SUMMARY AND FINDINGS

The information in this chapter does not constitute an assessment of all minerals critical to the U.S. economy. Rather, it demonstrates how such an assessment could be conducted given sufficient and appropriate information. In the process, the committee's assessment shows that designating a particular material as critical is a multifaceted and nuanced activity—the designation can differ from material to material, from country to country, and from user to user. This observation leads to the point that establishing the context for the evaluation of one or more minerals is important when employing the committee's matrix, or another methodology, to evaluate mineral criticality.

Minerals that rank high on both axes of the criticality matrix are characterized as "critical minerals," although it is important to understand that a mineral can rank high on one or both axes for quite different reasons. Examples discussed earlier in this chapter include the following:

- PGMs, several different applications of which are regarded as having high importance in use;
- REs, whose criticality is strongly dependent on supply risk concerns;
- Titanium, for which a minor use in terms of mass is vital to U.S. economic interests; and
- Lithium, which is not critical today, but could potentially become critical should a new use (hybrid vehicle batteries) be widely adopted.

In general, the committee found it easier to evaluate importance in use than risk to supply because there is no comprehensive, reliable, transparent, public evaluation of most of the aspects of supply risk. The U.S. government provides such evaluations for many considerations related to fossil fuel supplies; it would seem appropriate and useful for a similar set of evaluations related to nonrenewable but reusable resources to be performed as well.

The committee selected 11 mineral candidates for criticality analysis, based on committee members' own experiences and on minerals identified as potentially critical at the committee's March 2007 information-gathering meeting. Of these 11, 5 minerals or mineral groups—indium, manganese, niobium, PGMs, and REs—were determined to fall in or near the critical zone of the matrix (Figure 4.7). Although the United States is essentially completely dependent on imports for all five minerals or mineral groups, it is not import dependence per se that leads to the committee's determination that these are critical; rather in each case there are complementary circumstances that lead to significant supply risk, typically including a high degree of concentration of production in one or a small number

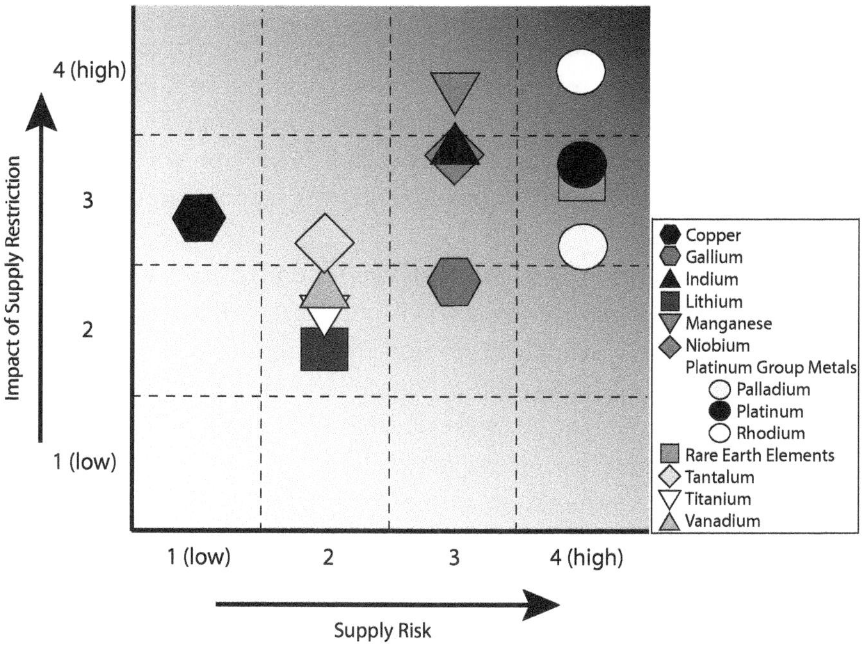

FIGURE 4.7 Criticality matrix for all 11 candidate materials discussed in the chapter. Of the 11, indium, manganese, niobium, PGMs, and REs fall in the critical "zone" of the matrix.

of countries or companies. However, the committee emphasizes again, as stated in Chapter 1, that the minerals analyzed in this chapter do not represent an absolute list of critical minerals. Other minerals could equally well be evaluated and determined to be critical for the nation, for individual industry sectors, or for other users. The 11 examples demonstrate the application of the methodology established by the committee for determining mineral criticality.

The committee makes the following findings on the basis of the example criticality assessments made in this chapter:

- The criticality matrix is a useful tool to evaluate the degree of criticality of a material. Criticality is not a state of being or not-being, but a location on a two-dimensional, multi-indicator continuum. Critical minerals are those that are both essential in use and subject to considerable supply risk.
- In placing a mineral or mineral product on the vertical axis (impact of supply restriction), technological importance in use is easier to evaluate than the other, largely economic factors. Technological importance in use depends primarily on whether or not technical substitutes for the mineral exist that can provide similar functionality.
- The economic impacts of a supply restriction depend on the degree to which costs rise if a mineral's supply is restricted and, more generally, on how a supply restriction affects a company's profitability (and in turn labor needs) or a nation's ability to supply a public good such as national defense. These economic impacts require case-by-case economic impact analyses.
- In placing a mineral or mineral product on the horizontal axis (supply risk), caution is required to assess supply risk because of the lack of suitable information on primary, secondary, and tertiary material flows and, more specifically, on subeconomic resources, by-product production, secondary production from scrap, intra-company trade, and mineral products embedded in imports and exports.

REFERENCES

Edwards, C.R., and A.J. Oliver, 2000. Uranium processing: A review of current methods and technology. *JOM* 15(9):12-20.

Herring, I. 2007. Presentation to the Committee on Critical Mineral Impacts on the U.S. Economy. Washington, D.C., March 7.

Johnson Matthey, 2007. Platinum 2007. Available online at *http://www.platinum.matthey.com/publications/Pt2007.html* (accessed August 9, 2007).

Joseph, G., 1999. *Copper: Its Trade, Manufacture, Use, and Environmental Status.* Materials Park, Oh.: ASM International.

Kohler, S., 2007. California. In state summaries. *Mining Engineering* (May):72-75.

Moore, P. (ed.), 2006. Regional review: South Africa. *Mining Magazine* (September):62.

Newton, F.T., S.P. Collings, and B.C. Little, 2006. Nuclear power update. *SEG Newsletter* (67):1, 8-15.

Stevens, L. 2007. Presentation to the Committee on Critical Mineral Impacts on the U.S. Economy. Washington, D.C., March 7.

USGS (U.S. Geological Survey), 1997. *Mineral Commodity Summaries 1997.* Washington, D.C.: U.S. Government Printing Office.

USGS, 2002. *Rare Earth Elements—Critical Resources for High Technology.* USGS Fact Sheet 087-02. Available online at *http://pubs.usgs.gov/fs/2002/fs087-02/fs087-02.pdf* (accessed November 13, 2007).

USGS, 2007. *Mineral Commodity Summaries 2007.* Reston, Va.: U.S. Geological Survey, 195 pp.

World Nuclear Association, 2005. Can Uranium Supplies Sustain the Global Nuclear Renaissance? Available online at *http://www.world-nuclear.org/reference/position_statements/uranium.html* (accessed November 13, 2007).

CHAPTER 5

Mineral Information and Possible Initiatives in Research and Education

INTRODUCTION

Minerals, as noted throughout this report, are important to the basic infrastructure of the nation, its productivity, and its economy. Therefore it is vital to evaluate every mineral's criticality with some regularity and with appropriate data collection and analysis so that decision making regarding critical minerals can occur in a time frame suitable to alleviate potentially disruptive restrictions on mineral availability. Because criticality is dynamic, the criticality of a mineral is really only a snapshot of a mineral's applications and of the risks to its supply at a given time; mineral criticality can be effectively evaluated only in the context of a continuous stream of baseline data and information.

This chapter discusses the need for the federal government to collect data on minerals, provide analysis of these data (including mineral criticality), and disseminate the data and analyses in a publicly accessible format. The chapter begins with a historical overview of the federal context for statistical data collection and mineral data collection, and follows with a catalogue of public and private mineral databases that can be used by policy makers, the public, and industry to make informed decisions about a mineral or group of minerals. With mineral criticality as a primary theme, the committee then suggests the types of information and research that could best enable informed decisions about mineral policies affecting the national economy and infrastructure. Finally, because the production and analysis of this information require trained professionals, the chapter surveys the state of education related to mineral resources.

MINERAL DATA AND THE FEDERAL STATISTICAL PROGRAM

The decision by the federal government to collect mineral data is founded on two themes: (1) public understanding of the importance of collection, analysis, and dissemination of statistical data and information about mineral use and demand, mineral production and supply, and other aspects of mineral markets; and (2) support at the highest levels of government for collection of mineral statistical data that address the full life cycles of minerals to inform and monitor public policy. Box 5.1 discusses more conceptually the justification for federal involvement in collection, analysis, and dissemination of mineral information and in mineral-related research.

Historical Perspective

The nation's historical commitment to mineral data collection has been robust, with support from both executive and legislative branches. Mineral data collection has been a recognized part of national policy since at least World War II (J. Morgan, Jr., personal communication, January 2007). During the past several decades, numerous pieces of legislation have affirmed the federal commitment to collect mineral information with a foundation in the importance of minerals to the national economy and national security. Among them, the National Materials and Minerals Policy, Research and Development Act of 1980 (P.L. 96-479) suggested that "the Executive Office shall coordinate the responsible departments and agencies to identify material needs and assist in the pursuit of measures that would assure the availability of materials critical to commerce, the economy, and national security." A report from the Subcommittee on Mines and Mining of the Committee on Interior and Insular Affairs of the U.S. House of Representatives (U.S. Congress, 1980, pp. 12-13) the same year emphasized further:

> There have been no less than 20 mineral or material policy studies that have been prepared or commissioned by one governmental agency or another, as well as others prepared for groups outside government. Additional studies have examined at least some part of mineral policy questions. Although many studies reflected some particular outlook or condition, all adopted as a universal starting point the national significance of adequate mineral supplies and the importance of a strong domestic industry. All agree, to a greater or lesser extent, that foreign imports provide least-cost benefits to the consumer. At the same time, most see the pitfalls of import dependency and how such dependency forfeits freedom to make political, economic, and defense decisions. *All strongly urge better governmental analytical capability and improved means of integrating information into a comprehensive picture portraying the synergistic impacts of governmental policies or actions upon industry and, beyond that, the broad national interest* [italics added]. In more recent reports, recycling and conservation receive more attention and a progressively stronger case is made for increased government participation in development of long range technological innovation.

A 1982 presidential report on minerals also noted the need to improve integration of mineral concerns into the policy process, suggesting that mineral information was useful to guide decisions on the strategic stockpile, land use, tax and tariff, trade, investment, research and development, and environmental protection, among others (Shanks, 1983).

Historically, then, a supportive federal view of the need to collect mineral information has been one part of the overall federal responsibility toward collection of various types of data for public use. As presented in previous chapters, the need for mineral data is no less important today than several decades ago. In fact, the need for more frequent and detailed

BOX 5.1
Mineral Information and Research: Why a Federal Role?

The need for information and research on minerals, by itself, does not automatically justify federal government activities. As noted by the National Research Council (NRC, 2003, p. 24):

> After all, in market economies there are natural and strong incentives for private entities producing and consuming minerals to carry out scientific research and to collect and disseminate information that is relevant and necessary for informed decision making. Nevertheless, in several specific circumstances, private markets are likely to yield suboptimal outcomes from the perspective of society as a whole.

Specific circumstances that are relevant when considering critical minerals include the following:

1. *The need for high-quality information.* Informed decisions require good information, whether the decisions are public (government) or private. An important and fundamental role of government in a market economy is to facilitate market activity. It does this in a number of ways, by establishing well-functioning legal and monetary systems and by facilitating the collection and dissemination of information necessary for informed decisions (e.g., economic data in the national income and product accounts). To be sure, there often are private sources of information, but these sometimes are proprietary or unavailable to the public. In other circumstances, the private sector does not collect specific types of information necessary for informed decisions, especially public policy decisions, for the reasons outlined below.
2. *Different time and risk preferences.* To the extent that private entities give less priority to the distant future than is optimal from society's perspective, the private market will underprovide information and research with benefits only in the distant future. Similarly, to the extent that private entities have a greater degree of risk aversion than society as a whole,

analyses of some critical minerals is at present more acute due to the highly global nature of the mineral market and increased global competition for mineral resources. As part of this commitment to federal collection of mineral data, very competent mineral data collection and public dissemination occur now through the efforts of the Minerals Information Team, overseen by the Mineral Resources Program (MRP) of the U.S. Geological Survey the private market will underprovide information and research whose benefits are risky or uncertain. Much of the mineral information and research that the committee identifies later in this report as appropriate for federal government involvement is beneficial only over the longer term or, in the case of research, is risky.

3. *Provision of public goods.* Economists distinguish between two types of goods, private and public. Private goods have the following characteristics: (1) one person's use of the goods reduces the amount available for someone else (e.g., if one person drinks a bottle of water, another person cannot drink it); and (2) it is easy to exclude people from benefiting from private goods if they do not pay for them (e.g., the convenience store will likely call the police if a person does not pay for bottled water). Most economists believe that these two characteristics of private goods mean that we generally can rely on private markets to supply these goods at appropriate prices and in appropriate quantities. Public goods, on the other hand, are likely to be underprovided by the private sector acting alone because of their characteristics: one person's use does not reduce the amount available for others; moreover, it is difficult to exclude someone who does not pay for public goods from the benefits provided by those goods. Consider national defense. Once it is provided, one person's benefit from it does not reduce the degree to which another person benefits, and if national defense were funded through voluntary contributions, it would be difficult to exclude someone who did not contribute to the funding of national defense from its benefits.

Mineral information and research are at least partially public goods. Once provided, many people and organizations can benefit without reducing other entities' benefits, even if they do not pay for the information or research. As a result, there is a crucial role for the federal government to play in facilitating the provision of those types of information and research that are largely public, rather than private, in nature. Such provision can be facilitated directly by public agencies, or by universities, or by other nongovernmental organizations.

SOURCE: NRC, 2003, pp. 24-28.

(USGS). These mineral data are used widely both by a variety of private concerns and throughout the federal government. However, the committee recognizes a difference in federal definitions of "principal statistical agencies," whose sole task it is to collect data and federal units such as the Minerals Information Team that are not designated principal statistical agencies but are nevertheless tasked to collect and disseminate data as a part of other mandated duties. The committee suggests that this difference in definition can affect the strength of a given data collection program as a function of resource allocation and overall program visibility and autonomy and discusses these issues below.

Federal Statistical Programs and Data Collection

The policy of the federal government toward the data and statistics it collects and disseminates explicitly acknowledges the role that good statistics play in informed decision making by both public and private sectors. A recent example of the federal policy approach to statistics is excerpted from the 2008 executive budget proposal, under the heading "Strengthening Federal Statistics" (OMB, 2007a, Section 4, p. 37):

> Federal statistical programs produce key information to inform public and private decision makers about a range of topics of interest, including the economy, the population, agriculture, crime, education, energy, the environment, health, science, and transportation. The ability of governments, businesses, and citizens to make appropriate decisions about budgets, employment, investments, taxes, and a host of other important matters depends critically on the ready availability of relevant, accurate, and timely Federal statistics.

The U.S. statistical programs are decentralized in nature. Of the more than 70 federal agencies or programs receiving funding to carry out statistical activities, only 13 are considered 'principal' federal statistical agencies and are members of the Interagency Council on Statistical Policy (ICSP). The other 60 or so agencies collect data in conjunction with other missions such as providing services or enforcing regulations. The 13 principal agencies follow (OMB, 2007a, Section 4, p. 38):

- Bureau of Economic Analysis (BEA), Department of Commerce
- Bureau of Justice Statistics, Department of Justice
- Bureau of Labor Statistics (BLS), Department of Labor
- Bureau of Transportation Statistics, Department of Transportation

- Census Bureau, Department of Commerce
- Energy Information Administration (EIA), Department of Energy
- Economic Research Service, Department of Agriculture
- National Agricultural Statistics Service, Department of Agriculture
- National Center for Education Statistics, Department of Education
- National Center for Health Statistics, Department of Health and Human Services
- Office of Research, Evaluation, and Statistics, Social Security Administration
- Statistics of Income, Internal Revenue Service, Department of the Treasury
- Science Resources Statistics Division, National Science Foundation

These principal statistical agencies received approximately $2.2 billion in FY 2006, which is about 40 percent of the funding for all agencies collecting statistical data (estimated to be $5.4 billion for FY 2007; OMB, 2007b).

Agencies or programs that are solely and directly responsible for statistical data collection are more likely to be able to develop and maintain an efficient program focused on the types of data necessary to support their missions. This ability is particularly important in order to meet the ICSP performance standards that directly address whether the unit performing the collection of data and distribution of information is effective. Those performance standards are used in completing the administration's Program Assessment Rating Tool (PART).

Conversely, federal units that have statistical data collection as only one feature of their body of tasks may not be able to maintain consistent resource levels and may lack broader federal program support necessary over the long term to achieve their data collection and dissemination goals or to maintain the important "effective" rating in PART. Furthermore, when a data collection unit is not designated as principal, its mission may be diluted or made subordinate to the overall mission of the agency or department of which it is a part, particularly in situations of constrained resources. From the standpoint of collecting data and statistics on minerals, the committee finds it significant that none of the 13 principal statistical agencies collects and publishes annual data on minerals and their availability. As previous chapters have demonstrated, minerals are essential natural resources that are intrinsic to the availability and function of an enormous number of consumer goods in the nation and to the general national infrastructure.

Lacking somewhat the direct macroeconomic impact that energy has on the nation, minerals nonetheless factor into the overall function and productivity of the country. Critical minerals, in particular, may play very specific roles in the development of different industry sectors and their products. In this regard, the committee found a potential analogue for the concept of principal mineral statistical data collection through the example of the EIA. Box 5.2 reviews the history, purview, and federal role of the EIA, together with its impact on energy policy, due in part, in the committee's view, to its status as a principal statistical agency.

Confidentiality of Federal Survey Information

An understanding of the role of data confidentiality in all federal data collection activities is important to acquire a fuller understanding of the basis for federal engagement in collecting data for policy formulation and implementation. Confidentiality of proprietary data by privately and publicly owned companies is considered to be a fundamental right, and for some a necessity, in a world economy subject to intense competition. Recent affirmation of the need to balance federal data collection with respect for privacy was put forward in the language of the 2008 executive budget proposal (OMB, 2007b, Section 4, p. 37):

> As the collectors and providers of these basic statistics, the responsible agencies act as data stewards—balancing public and private

decision makers' needs for information with legal and ethical obligations to minimize reporting burden, respect respondents' privacy, and protect the confidentiality of the data provided to the Government.

The need to protect the confidentiality of data provided to the federal government and the need to respect respondents' privacy suggests that the primary mineral data made available to the public (and indeed all fundamental data gathering) be largely conducted in an organized and accountable manner by federal government programs or agencies specifically tasked and qualified to do so; federal programs can collect such data in an unbiased way that ensures the anonymity of the respondents and subsequent use of the data for the public good. Collection of data by the government does not preclude their collection by various nongovernmental entities, including those who do so to satisfy a commercial need for data not collected by government. When nongovernmental entities do collect information, they usually do so in order to satisfy a specific constituency and not necessarily the needs of government to formulate and monitor public policy. Importantly also, even if confidentiality concerns could be guaranteed, the cost of thorough and accurate data collection at a national scale is high and may be excessive for a nongovernmental entity compared to the market value of the data.

The principle of data confidentiality adhered to by federal statistical agencies and programs is based on the belief that individuals and businesses will be more likely to give complete and factual answers to surveys when they are assured that their responses will not be used to bring legal or regulatory actions against them. In some instances, particularly where businesses are in highly competitive markets, respondents may fear that disclosure of certain business data to competitors or even customers will result in damage to their business. It is believed that businesses will be more likely to trust government agencies to maintain confidentiality than an association or other nongovernmental entity.

Finally, in terms of contrasting a strong governmental statistical role in collection of mineral or other data with the role played by nongovernmen-

BOX 5.2
Domestic Energy Information: An Analogue for Mineral Information Collection

The oil embargo of 1973 and resulting fluctuations in the nation's supply and price of oil prompted a congressional response in 1974 through passage of the Federal Energy Administration (FEA) Act. FEA was the first federal agency with a primary responsibility for energy issues, and a key FEA function was the collection, analysis, and dissemination of national and international energy information (U.S. Congress, 1974). As part of its energy information collection procedures, FEA was given data collection responsibility with related enforcement authority for its energy surveys of energy suppliers and major energy consumers. Energy data were to be used for statistical and forecasting activities and analyses such as the structure of the energy supply system, including resource consumption; energy resource reserves, exploration, development, production, transportation, and consumption sensitivities to economic factors, environmental constraints, technological improvements, and substitutability of alternate energy sources; and long-term industrial, labor, and regional impacts of changes in patterns of energy supply and consumption.

In 1977, the Department of Energy (DOE) replaced the FEA as set forth in Public Law 95-91. The EIA was established as part of DOE to incorporate the information-gathering tasks and functions described for FEA, in addition to the responsibility ". . . for carrying out a central, comprehensive, and unified energy data and information program which will collect, evaluate, assemble, analyze, and disseminate data and information which is relevant to energy resource reserves, energy production, demand, and technology, and related economic and statistical information, or which is relevant to the adequacy of energy resources to meet the demands in the near and longer term future for the Nation's economic and social needs" (U.S. Congress, 1977). The security of the nation's energy supply was thus a concept embedded in this public charter.

Although located within and working in close collaboration with DOE, the EIA maintains independence with regard to its work. This concept of independence is embodied in its enabling legislation, mandating that the EIA administrator "*shall not be required to obtain the approval of any other officer or employee of the Department [of Energy] in connection with the collection or analysis of any information; nor shall the Administrator be required, prior to publication, to obtain the approval of any other officer or employee of the United States with respect to the substance of any statistical or forecasting technical reports which he [she] has prepared in accordance with law*" (U.S. Congress, 1977).

With a focus on gathering, analyzing, and forecasting energy information for the public and private sectors, the EIA's four program offices cover oil and natural gas; coal, nuclear, electric, and alternate fuels; energy markets and end use; and integrated analysis and forecasting. Four support offices provide the assistance required for effective and efficient dissemination of energy information. EIA works to meet the needs of its congressional clients as well as other important customers including state and local governments, federal agencies, industry, and the general public. Particularly interesting to many clients is the product and service line that incorporates time series data based on EIA's 65 recurring statistical surveys. Other EIA data sources include international information exchanges with individual countries or agencies such as the International Energy Agency. The relationship between the EIA, industry, and other public and private organizations is symbiotic: industry and other organizations rely on the data, analyses, and forecasts disseminated by EIA, and EIA relies on thorough and accurate information from these entities to amalgamate into its national energy information system.

The interrelationships of energy, the economy, the environment, and the quality of life of U.S. citizens have led to significant increases in demands for EIA's information. For example, since 2000 the need for petroleum information has increased as evidenced by greater public demand and inquiries, as well as congressional requests, for information. As a result, the EIA made a decision to elevate the quantity, frequency, and types of petroleum information available to the public. For example, "This Week in Petroleum," available online at *http://tonto.eia.doe.gov/oog/info/twip/twip.asp*, was introduced in 2002 to improve public understanding of EIA's petroleum data and the many factors affecting petroleum markets. In the committee's parlance, oil is considered a critical fuel type by the EIA, and the amount and frequency of data collected for this fuel type were increased recently to satisfy public and private demands to plan for and react to potential fluctuations in the petroleum supply.

Importantly, although greater need for information may dictate an increase in the frequency of data reporting for distinct fuel types, baseline data collection is always maintained on the other fuel types in EIA's portfolio. Gaps are difficult or impossible to fill if data collection on a fuel type ceases. Furthermore, while baseline data are important all the time, baseline data collection is most important—yet most difficult—during times of crisis, underscoring the need to maintain good baseline data sources to weather any crisis period.

As an independent, federal entity, EIA data are presented in an unbiased manner where statistical and quality standards are followed. Data breakdowns and presentations include full data sets at national levels, by sector or state, and by industry, as well as analyses and forecasts. EIA's analyses and forecasts are interpretations or models of data, and these interpretations or models are only as credible or defensible as the data upon which

continued

BOX 5.2
Continued

they are based. Errors in any data presented by the EIA may result in inappropriate market responses. Industry cooperation in accurately completing data surveys is thus paramount to maintaining the quality of EIA's data sets. Although industry has a very high rate of complying with its surveys, EIA benefits from the legally mandated collection enforcement authority it possesses, as well as the provisions it makes with regard to protecting sensitive information. The committee notes that recent legal proceedings surrounding the maintenance of data confidentiality in certain federal investigations have raised doubt about federal agencies' ability to protect data sources in all situations (Anderson and Seltzer, 2005).

Accurate and thorough data collection and analysis are time-consuming and expensive processes, and the EIA must continually demonstrate the benefit to the public of collecting the data. Interesting, useful, and timely data presentation with strong basis in effective dissemination is one of the keys to this public education process.

tal entities that also collect such data, consideration must be given to the fact that free public data availability is a key feature of making informed policy decisions and of fulfilling a responsibility to the public served by the federal government. Despite comprehensive and often very useful statistical gathering, analysis, forecasting, and publications by nongovernmental entities, usually limited information is available free to the public from such sources simply because these entities must recover the cost of data collection and analysis; thus, more comprehensive data, analyses, forecasts, and publications are often available only on payment of a subscription fee. The federal government can and does purchase these subscriptions for specific purposes, but it must still collate and analyze the data and cannot always document the data sources or the degree of duplication between different data sets. Data purchase from private sources by the federal government is necessary and useful for a variety of purposes, but it is not considered adequate to serve identified public policy needs.

Current federal mineral data collection through the USGS Minerals Information Team follows many of the same basic procedures as does the EIA, particularly with regard to attempting to ensure accurate and timely mineral data collection and the general scope of its mandate, but with two important differences: (1) the Minerals Information Team is not a defined, principal statistical agency and as such is not autonomous either in function or in resource allocation from the Department of the Interior and USGS; and (2) though it does largely receive good cooperation from the private sector, the Minerals Information Team does not have collection enforcement authority for its surveys, nor did its predecessor, the U.S. Bureau of Mines.

SOURCE: EIA, 2006; DeYoung (2007); M. Kaas, personal communication, June 2007; J. Shore, personal communication, March 2007.

CRITICAL MINERAL INFORMATION SOURCES

Information about minerals produced and used in the United States is available from a wide variety of public sources and nonprofit or for-profit organizations. These sources are based both in this country and abroad. In this section the committee describes some of these information sources, with the caveat that the list is by no means exhaustive.

Many mineral data sources generate limited information or produce data on an ad hoc basis, rather than as part of a regularly scheduled program. Doctoral and master's theses and other academic and trade publications can contain valuable data and insightful analysis, but distribution is often limited and may not be timely (too early or too late) to be useful in determining criticality, so these sources are not discussed further here. In addition, numerous organizations have Internet web sites dedicated to providing information either in support of or against mining and its effects on the environment and local communities; despite the potential for bias, these sites often can present valuable information to assist in developing appropriate policy responses to address critical minerals, but again these sources were not considered in the committee's analysis because of the difficulty in documenting the background protocol used to collect and collate the data.

The committee examined numerous international government data sources, domestic and international nongovernmental data sources, and U.S. federal data sources of mineral information (see also Table 5.1). Of the international government organizations collecting such data, geological surveys are often rich sources of mineral information. The information is often focused on the survey's home nation, but very often these databases may provide international data as well. As one example, Box 5.3 provides a description of the Japanese Oil, Gas and Metals National Corporation (JOGMEC) and the role it plays in data gathering, analysis, and information dissemination in a nation that is heavily dependent on imports for its minerals. The committee found that almost all nations, and sometimes associations of nations, have mineral information in one form or another, but the quality and quantity of the data and their presentation and accessibility are highly variable. The committee notes that not all foreign geological surveys allow free public access to their assembled data sets.

Numerous private organizations collect and sell mineral data and related reports targeted to specific business needs. Information from these sources has the potential to fill gaps that exist in government statistical products, analyses, and forecasts. Additional information is available from nongovernmental professional associations and organizations that collect data about their respective industries and make some data available to the public or to subscribers or members on a fee basis. Nearly every major metal and many minerals are represented by one or more dedicated professional associations, all of which seek to provide market information to their constituencies and the public. The committee does note these information sources as potentially very useful supplemental sources of specific information on single minerals or specific groups of minerals. They carry with them, however, the difficulty in documenting data sources, data collection protocol, and potential overlap with other data sets from competing

TABLE 5.1 International and Domestic Sources of Mineral information

Classification	Name	Internet Sites
International	British Geological Survey	*http://www.bgs.ac.uk/enquiries/mins.html*
	Natural Resources Canada	*http://www.nrcan.gc.ca/com/index-eng.php*
	Australia's Commonwealth Scientific and Industrial Research Organization	*http://www.csiro.au/*
	Bureau de Recherches Géologiques et Minières	*http://www.brgm.fr/*
	Japanese Oil, Gas and Metals National Corporation	*http://www.jogmec.go.jp*
	Eurostat	*http://epp.eurostat.ec.europa.eu/*
	Mineral Resources Forum of the United Nations Conference on Trade and Development	*http://www.natural-resources.org/minerals/*

continued

TABLE 5.1 Continued

Classification	Name	Internet Sites
Private and nongovernmental	Port Import Export Reporting Service	*http://www.piers.com/*
	Metals Economics Group	*http://www.metalseconomics.com/default.htm*
	Johnson Matthey	*http://www.matthey.com/*
	National Mining Association	*http://www.nma.org/*
	Portland Cement Association	*http://www.cement.org/*
	Aluminum Association, Inc.	*http://www.aluminum.org/*
	Copper Development Association	*http:/www.copper.org/*
	Nickel Institute	*http://www.nickelinstitute.org/*
	American Iron and Steel Institute	*http://www.steel.org/*
	Great Western Minerals Group, Limited	*http://www.gwmg.ca/about_us/index.php*
	Minerals Information Institute	*http://www.mii.org/*
	International Copper Study Group	*http://www.icsg.org/*
	International Nickel Study Group	*http://www.insg.org/*
	International Lead And Zinc Study Group	*http://www.ilzsg.org/*

U.S. federal	Minerals Information Team of the USGS Mineral Resources Program	*http://minerals.usgs.gov/minerals/*
	Department of Commerce, U.S. Bureau of the Census	*http://www.census.gov/epcd/www/naics.html*
	Department of Commerce, Bureau of Economic Analysis (BEA)	*http://www.bea.gov/index.htm*
	Department of Labor, International Trade Administration	*http://trade.gov/index.asp*
	Department of Labor, Bureau of Labor Statistics (BLS)	*http://www.bls.gov/*
	Department of Labor, Mine Safety and Health Administration	*http://www.msha.gov/*
	Department of the Interior, Bureau of Land Management	*http://www.blm.gov/*
	Department of the Interior, Office of Surface Mining	*http://www.osmre.gov/*
	Department of the Interior, Minerals Management Service	*http://www.mms.gov/*
	Department of Agriculture, U.S. Forest Service	*http://www.fs.fed.us/*
	Department of Health and Human Services, Centers for Disease Control and Prevention, National Institute for Occupational Safety and Health	*http://cdc.gov/niosh/*
	Department of Defense, National Defense Stockpile	*https://www.dnsc.dla.mil/default.asp*

BOX 5.3
Japanese Oil, Gas and Metals National Corporation

Japanese industry is highly reliant on imported minerals and mineral products. The Japanese government, for a number of years, has undertaken activities to facilitate stable supplies. In 2004, it established the Japanese Oil, Gas and Metals National Corporation (JOGMEC) to undertake these activities, integrating activities previously carried out in the Japan National Oil Corporation (since 1967) and the Metal Mining Agency of Japan (since 1963). JOGMEC is responsible for oil, natural gas, and metals and minerals. For metals and minerals, among JOGMEC's important activities are providing financial assistance to Japanese companies for mineral exploration and deposit development, gathering and analyzing information on mineral and metal markets to better understand supply risk, and managing Japan's economic stockpile for rare metals. JOGMEC defines rare metals as those that (1) are essential to Japanese industry—sectors such as iron and steel, automobiles, information technology, and home appliances—and (2) are subject to significant supply instability.

JOGMEC gathers, analyzes, and disseminates information to assist Japanese companies and government agencies. Types of data and information include overseas geology and ore deposit descriptions and interpretations, mineral policies and regulations for other nations, market data and analysis (supply, demand, prices, etc.), and information on mining and the environment.

JOGMEC manages rare metal stockpiles in cooperation with private companies. The goal is to have stocks equivalent to 60 days of Japanese industrial consumption—42 held by JOGMEC and 18 by private Japanese companies. At present, stocks exist for seven metals: chromium, cobalt, manganese, molybdenum, nickel, tungsten, and vanadium. JOGMEC is closely observing seven other metals or groups of metals: gallium, indium, niobium, platinum, rare earths, strontium, and tantalum. A set of guidelines and rules govern release of metals from stocks. In 2005, the stockpile released nickel and tungsten to the market.

NOTE: Chapter 4 evaluates 8 of the 14 metals JOGMEC either stockpiles or is evaluating closely. Stated slightly differently, 8 of the 11 minerals evaluated in Chapter 4 are being stockpiled or monitored by JOGMEC.

SOURCE: JOGMEC, available online at *http://www.jogmec.go.jp* (accessed June 21, 2007); Murakami, 2007.

associations or private organizations. For many of these data assemblages, a collating and analysis activity is required if the data are to be of maximum utility. This activity is scarce or absent within the federal government at present.

For the federal government, the Bureau of the Census of the Department of Commerce conducts a Census of the Mineral Industries every 5 years, for years ending in 2 and 7. All domestic mining establishments with at least one employee are surveyed. The content of the census includes the kind of business, geographic location, legal form of ownership, total revenue, annual and first-quarter payroll, and number of employees for the pay period including March 12. Receipt of a "long form" requires the establishment to provide added detail on employment, payrolls, worker hours and payroll supplements, value of shipments, inventories, capital expenditures, and quantity and value of products, supplies, and fuels. A short form requests only basic data in the form of capital expenditures and quantity and value of products.

While participation in the census is mandatory, the data collected provide a limited picture of specific minerals due both to reports that withhold information to avoid revealing confidential data and to the aggregation of categories. The data categories are useful for providing input to national income and product accounts, but the types of data collected are insufficient for detailed life-cycle analysis or for determining criticality because information on production and consumption of most mineral products is not collected or reported.

The Department of Commerce BEA and International Trade Administration are important users, rather than collectors, of mineral data in the conduct of their policy discussions and in assessing the effects of trade, tariff, and nontariff barriers, regulation, licensing schemes, and international competition on the function of commerce in the United States (Cammarota, 2007). In the federal structure, these agencies' policies are supported by the data and information they are able to collate from other sources. For mineral information, the Department of Commerce relies heavily on the USGS Minerals Information Team, but also uses other data sources, including information from federal study groups and industry associations. Disaggregated mineral data were noted as a preference for the Department of Commerce whenever possible.

Mineral information is also collected by other government agencies on a basis that is usually quite specific to the mission of the agency or a defined, finite project or task. Examples are the Bureau of Land Management, the Mine Safety and Health Administration, the Office of Surface Mining, the Minerals Management Service, the U.S. Forest Service, the National Institute for Occupational Safety and Health, and the Department of Defense (DoD) National Defense Stockpile. Because the data are usually collected for a specific purpose, they are often not made easily available to the public and may not be published in a form that allows simple comparison with data from other government sources.

The most comprehensive source of mineral information for the United States is the Minerals Information Team of the USGS MRP (Figure 5.1). The Minerals Information Team is the sole agency within the federal government tasked to "collect, analyze, and disseminate information on the domestic and international supply of and demand for minerals and mineral materials essential to the U.S. economy and national security" (USGS, 2007). In addition to traditional mineral production data, Minerals Information Team analyses include some amount of information and reports on mineral conservation, sustainability, material flows, availability, and the economic health of the U.S. mineral industry.

To produce its assessments and reports, the group requests data on mineral production, consumption, recycling, stocks, and shipments from the mining and mineral processing industries in the United States; more than 140 surveys are conducted annually on commodities. Aggregated U.S. statistics are published to preserve proprietary company data. The Minerals Information Team does not have enforcement authority with these industry surveys (see also Box 5.1). Other data include those collected annually from more than 18,000 voluntary producer and consumer establishments that complete about 40,000 survey forms. The USGS also has cooperative agreements with state governments to exchange data. Data from other federal agencies, primarily the Department of Commerce, are also included in the analyses.

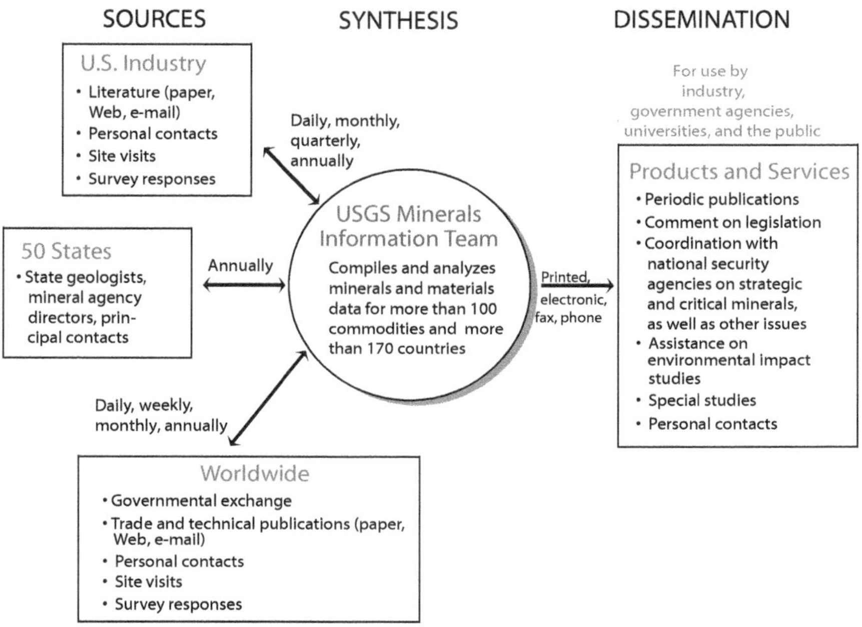

FIGURE 5.1 Diagram showing the main activities, data sources, partnerships, and work flow for the USGS Minerals Information Team. SOURCE: USGS, 2007.

The Minerals Information Team's customers include the general public, Congress, industry, and other federal agencies, including the Department of Commerce, the DoD National Defense Stockpile, the Federal Reserve Board, the Bureau of Land Management, the U.S. Forest Service, and the Environmental Protection Agency. The committee was struck by the uniform responses it received from industry, government, and academic participants at its open meeting sessions (Appendix B) and in voluntary written contributions to its work (e.g., Ellis, 2007) that USGS Minerals Information Team data were the primary data to which they turned for immediate information on minerals in their respective fields. Nonetheless, the Minerals Information Team activities are not as comprehensive as is desirable. Reflecting the history and focus of the parent agency, Minerals

Information Team analyses do not regularly address in-use stocks of metals, scrap stock and trade, detailed recycling information, and other aspects of mineral availability and criticality centered on stages of the material life cycle well removed from mineral extraction and processing.

ENHANCING OUR UNDERSTANDING OF CRITICAL MINERALS: FEDERAL INFORMATION AND RESEARCH NEEDS

The Mineral Cycle, Federal Information Exchange, and Research

As previous chapters have demonstrated, minerals are rarely used in their raw form as an end product. They are incorporated into an intermediate or final product that takes advantage of the properties of the particular mineral to improve the performance of the final application. For that reason it is not enough to have information only about the primary production of the mineral. As outlined in Chapter 1, the entire life cycle of the mineral is important because it directly affects the supply-demand relationship. If markets function well, the price mechanism will bring about adjustments in the supply-demand relationship that will restore the balance over time. If the market mechanism is to function properly it is absolutely essential that the market have good information about both sides of the equation. Good information about all stages of the life cycle of the mineral becomes increasingly important if markets are not completely well functioning—for example, if significant environmental damage accompanies production or if innovative research is an important determinant of long-term mineral availability.

The USGS MRP, which has oversight of the Minerals Information Team, is the only federal program that provides resource assessments and research on mineral potential, production, consumption, and environmental impacts. The committee is supportive of the incorporation of critical minerals as a specific piece of this type of mineral life-cycle analysis by the MRP and the Minerals Information Team and by other federal agencies collecting mineral-related data.

The most recent attempt by Congress to ensure that all federal statistical agencies would be able to protect the confidentiality of data collected by them is the 2002 Confidential Information Protection and Statistical Efficiency Act (CIPSEA). In addition to data confidentiality, other provisions in CIPSEA are also important with respect to improvement in all federal data collection. CIPSEA legislation has provisions intended "to promote efficiency in the production of the nation's statistics by authorizing limited sharing of business data for statistical purposes. The objectives behind the data-sharing component of the legislation are threefold. First, it was hoped that permitting the three agencies (Census Bureau, BEA, and BLS) to share information would improve the comparability and accuracy of federal statistics by allowing timelier updates of sample frames, development of consistent classifications of establishments and industries, and exploitation of administrative data. Second, more integrated use of data should reduce the paperwork burden for surveyed businesses. Finally, through these mechanisms, it was hoped that the sharing of data would lead to improved understanding of the U.S. economy, especially for key industry and regional statistics" (NRC, 2006). The committee agrees that sharing statistics and the survey lists used to collect data will lead to better, more complete information and will minimize duplication of data sets, facilitating overall improvement in federal mineral and economic analysis. At present, the Minerals Information Team is not officially included in the sharing of survey lists, so it is difficult to compare and collate data collected and published by the Census Bureau, the BLS, or the BEA with data collected and published by the Minerals Information Team.

To develop a good understanding of both the criticality and the complete life cycle of one or more minerals, information collected in the following areas, in addition to the standard primary production and demand for each mineral, could be useful:

- Recycling or scrap generation and inventories of old scrap
- In-use stocks
- Reserves and resources
- Downstream uses

- Subeconomic resources
- Material flows
- International information in all of the above areas

There currently is no source of information that supplies all of the above data on at least an annual basis for all minerals.

In a number of areas relevant to critical materials, the committee found that there was a paucity of information or that appropriate technology had not been developed or pushed very far. The committee relates these deficiencies to a low level of support for research on resource availability and resource technology. In particular, the following topics require substantially increased research effort if critical materials are to be identified reliably in the future, the sources of those materials (from both virgin and recyclable stocks) are to be better quantified, and the technology for extraction and processing is to be substantially enhanced:

- Theoretical geochemical research to better identify and quantify virgin stocks that are potentially minable;
- Extraction and processing technology to improve energy efficiency, decrease water use, and enhance material separation;
- Remanufacturing and recycling technology, which is the key component in increasing the rate and efficiency of material reuse; and
- The characterization of stocks and flows of materials, especially import and export, as components of products, and losses upon product discard. This lack of information impedes planning on many levels.

A Federal Model to Address Critical Minerals as Part of Public Policy

Once mineral criticality has been established, the question arises as to what public policy actions could circumvent or eliminate the factors that have brought about the critical nature of the mineral. The Industrial College of the Armed Forces (ICAF) National Defense University issued a report on strategic materials in 2006 that addressed the question of what public policy ought to do toward ensuring the availability of strategic minerals (ICAF, 2006). The public policy issues are similar for both strategic minerals and critical minerals, and the committee finds the ICAF analysis to be a structured and pragmatic way to view critical mineral data collection and organization. A brief review of this analysis is offered here as one potential, published, federally oriented perspective by which a federal mineral information collection and research program could be enhanced and strengthened to address critical minerals. The military is clearly only one of the many customers for mineral information, and its suggested approach has practical merits.

Among the general findings in the analysis, two are especially relevant and applicable to critical minerals. First, the report stated that "more effective collaboration between industry, government and academia could improve the industry's ability to cope" (ICAF, 2006, p. 20) with the various challenges facing suppliers of strategic materials. As an example, the study refers to the potential for collaboration to streamline legislative, regulatory, and policy issues. Second, the report stated that "increased support for education and research throughout American society would provide for continued innovation in materials science" (ICAF, 2006, p. 20).

The report also contains a number of specific findings organized around strategic materials at different stages in the product life cycle. At the *mature* stage, the report suggests that materials typically are subject to increased international competition and that companies primarily pursue strategies aimed at operational efficiency and low costs. The report cites steel and aluminum as examples. The report also argues that it is important for governments to avoid inappropriate protection for mature industries, except in exceptionally limited and compelling circumstances.

For materials experiencing significant *growth*, on the other hand, key challenges are gaining access to raw materials when demand is increasing rapidly and facilitating rapid transfer of new technologies. The report cites advanced composites, superalloys, and specialized materials using rare earth elements as examples (ICAF, 2006). A key role for government here is in facilitating and enabling technology transitions and in monitoring markets for these growth materials for potential supply restrictions. The recommended role of government regarding minerals in growth markets reflects the fact that many materials in new technological applications have come about through government involvement in research and development to achieve higher performance in new applications. Oftentimes military use alone is insufficient to achieve economic viability of a material without commercial applications as well. Government therefore needs to foster applications that will improve the commercial success of the material in order to ensure its availability at reasonable cost for military applications. One focus in this report has been to put into place a framework for determining mineral criticality, some of which are in a growth phase of development according to definitions presented in the ICAF (2006) report.

Finally, the report argues that government's most significant role relates to *breakthrough* materials requiring significant research and development that is high risk and has uncertain rewards. As examples of breakthrough technologies and materials, it cites nanotechnologies, microelectromechanical systems, and biometrics. The "ventures in breakthrough technologies face development difficulties that require substantial R&D and engineering. Within the materials industry, specific problems include long development cycles, low levels of funding, stakeholder concerns, and foreign competition. Overcoming these challenges merits greater government attention than typically required for more mature materials markets" (ICAF, 2006, p. 10).

The ICAF report recommended that the role of government regarding minerals in breakthrough markets should be to provide active support for the development of the next generation of materials. Experts in the materials field view DoD as a key federal organization to coordinate and fund materials research, recommending additional research into the discovery and characterization of materials with unique or substantially improved properties (NRC, 2003). Defense leaders acknowledge this key role because "many emerging defense suppliers find it difficult to raise funds for military R&D and project opportunities" (DUSD(IP), 2003, p. B-8). While many stakeholders would welcome increased funding for advanced materials research, a fiscally responsible initial step could prioritize materials research endeavors to ensure that the nation's most important requirements are met.

Many government efforts specifically focus on innovative research in materials specialties. These efforts support a variety of worthwhile research in materials science. However, individual agencies award many of these grants on an individual or somewhat ad hoc basis that is not the product of a coordinated research strategy. In particular, they rarely address mineral information needs or consider mineral supply and demand data or criticality, either short or long term.

The ICAF (2006) study is silent about the important question regarding the source or sources from which the information will come to conduct the recommended analysis and support the resulting mineral cycle categorization. Studies going back many years have repeatedly made the case for a federally supported and funded program to collect and disseminate the mineral data necessary to make good policy decisions. The committee views materials research as an integrated part of the support for and use of high-quality mineral information and research; indeed, materials and mineral research complement one another, with mineral research and data collection providing information for breakthroughs in new materials; these breakthroughs provide new paths along which to view the dynamics of mineral criticality.

THE PROFESSIONAL PIPELINE

Highlighting the need for an innovative and educated workforce to ensure the nation's continued growth and prosperity, the NRC (2007) report *Rising Above the Gathering Storm* drew attention to the decreasing number of students being educated in all science and engineering fields in the United States and raised questions about the potential detrimental effect of this trend on future domestic technological innovation, manufacturing productivity, and general economic well-being. Coupled with demographic trends that show a large, aging workforce and a disproportionately smaller number of younger persons able to replace these professionals as they retire (The Economist, 2006), the ability of the nation to respond to surges in demand for qualified workers in industry, the federal government, and educational fields is suggested to be hampered if changes are not enacted in the national approach toward science and engineering education at all levels.

While the demographic issue is a global phenomenon, mineral availability over the longer term—in terms of both quantity and quality of resources—as discussed in this report depends importantly on specific types of resource professionals. More specifically, well-educated resource professionals are essential in industry, the federal government, and academic institutions for fostering the innovation that is necessary to ensure the availability of critical mineral resources at acceptable costs and with minimal environmental damage. These professionals include persons trained in what may be considered the more traditional fields of economic geology, mining engineering, and mineral processing engineering as well as persons in related specialties of resource analysis, resource economics, and environmental engineering. In the most recent decade, realization of the existence and magnitude of secondary resources in potentially recycled materials has added a need for two types of specialists not heretofore recognized or supported: product design engineers specializing in "design for recycling" and recycling technology engineers. By and large, only a handful of these "end-of-life" specialists exist worldwide, and those who exist were trained on the job rather than in dedicated academic programs.

With the collapse of the job market for undergraduate and graduate students in the 1980s and 1990s in fields related to the mineral engineering disciplines, schools and programs in these fields have disappeared relatively quickly. In addition to the lack of mineral and mining engineers in training at universities and colleges, very few formal training programs exist in materials recycling technology. While a number of highly qualified graduates in the general engineering disciplines, including environmental engineering, are being employed and making strong contributions to the mineral and mining industry, the skills and training for specific types of work in mining engineering, mineral economics, metallurgy, and recycling, for example, are not necessarily easily transferred, and on-the-job retraining between disciplines can be very demanding in terms of resources and time.

The private sector has underscored the need now for new graduates and midcareer persons with mining, geoscience, and/or engineering backgrounds to fill open positions in its organizations in all areas of research, exploration, production, health and safety, and environmental regulation and compliance (e.g., LeVier, 2007). Implicit in this report with regard to critical minerals is the added need to maintain adequate, accurate, and timely information and analysis on minerals at a national level in the federal government with additional, not fewer, professionals having appropriate backgrounds to perform the work. The committee did not conduct a formal survey of numbers of professionals with mineral- or material-related expertise in all federal agencies. Statistics available from the National Science Foundation (NSF, 2005) on the general employment trends in science and engineering in the federal government between the years 1998 and 2002 indicate that in the fields of geology, geophysics, mining engineering, metallurgy, or more generally natural resource operations, the number of federal employees stayed roughly the same in all fields except metallurgy and mining engineering. In the latter two fields, a slight decrease was recorded across the entire federal government. These statistics did not provide, and were not designed to provide, a detailed breakdown within specific departments or offices of each agency, nor did they provide an indication of the types of work the individuals in these fields perform or their age demographic. The committee thus acknowledges and views with concern the lack of supply of persons with appropriate backgrounds in mineral- and material-related fields noted during the course of this study in discussions with the DoD, Department of Commerce, and Department of the Interior.

For example, several years ago the Department of Commerce standardized all of its position descriptions within its industry offices. The department made a concerted effort to hire "generalists" who could easily be assigned to several industries or reassigned between industries. As a result, specific industry experience or expertise was eliminated. In addition, specific industry coverage was reduced: during the past 20 years the Department of Commerce eliminated the Copper, Lead/Zinc, Aluminum, Steel, and Miscellaneous Metals Divisions. These divisions were combined into one Metals Division. Currently only four individuals are responsible for issues affecting metallic raw materials.

Historically also, the U.S. Bureau of Mines (USBM), because of its mandate, had the highest concentration of mining-related professionals in the federal government. At one point, more than 1000 individuals, including project and support staff, were employed at the USBM. The closing of USBM in 1995 curtailed or terminated much of the government's research activities related to mineral resources (M. Kaas, personal communication, March 2007), although data acquisition and dissemination remains as a high-quality but (the committee believes) underfunded program now at the Minerals Information Team within the USGS. Natural attrition and funding issues such as these, as well as administrative decisions on the part of the U.S. government, have compounded the existing demographic trend for groups like the Minerals Information Team. The organization has seen a decrease in staff levels of 29 percent since 1996, including a variety of mineral specialists, some of whom also contribute language interpretation skills for international mineral data in their global analyses. Thus, while the U.S. government is itself, to a large degree, responsible for its current staffing situation regarding mineral-related professionals, the fact that numerous federal agencies collect, use, and analyze mineral and mineral economic data, conduct mineral-related leasing or regulatory activities, work in mining engineering and mine safety projects, or perform mineral- and material-related research and analysis suggests that the number of professionals in these fields has to be maintained or increased, not reduced, as natural attrition through retirement begins to take place given the demographics at these agencies.

Although it is tempting to extrapolate these observations to private industry and academic sectors, as outlined more generally in the National Research Council report (2007), the committee did not have the time or purview to establish labor and training statistics in all mineral-related fields across the private, public, and academic sectors to make a broader statement, and that work is left for another study to assess in detail. However, the committee would urge that implementing the conclusions and recommendations of this study occur simultaneously with examination of the workforce situation, including the appropriate number and skills of professionals needed and available to conduct effective work related to information gathering and research on critical minerals in a global, material-flow framework.

The committee's opinion is that present student levels and retirement trends at universities in several key mineral-related fields question the capacity to meet future workforce requirements. Market responses may eventually cover some of the apparent gap between demand for workers, and the supply of new hires to fill open positions in mineral-related fields, but the time lag between the beginning of the market response and the actual entry of educated and trained persons into these positions entails larger commitments than the market alone is able to address.

SUMMARY AND FINDINGS

This chapter has provided past legislative and executive references that have demonstrated support for comprehensive federal data collection as an integral part of both national policy and, specifically, national mineral policy. In its examination of federal statistical programs, the committee has found that while the total federal commitment to data collection is evident, the detail and depth to which a statistical agency or program can collect and disseminate *accurate and timely* data may depend on the agency or program's autonomy, the resources it is allocated as a function of that autonomy, and the authority with which it is allowed to enforce collection of survey data. Federal mineral data collection does not currently have the same authority given to designated federal statistical agencies and, by that omission, may lack some of the authority this committee finds to be appropriate for the important task of collecting mineral information and its incorporation in decision making that affects the U.S. economy. In its functions and legislated authority, the EIA is provided as a potential example by which current federal minerals information collection could be strengthened and made more effective; this includes enforcement authority for mineral survey responses and the autonomy to make professional decisions regarding publication style and frequency for critical minerals.

In terms of sources for mineral data, information, and analysis, a large range of international, private, nongovernmental, and federal organizations or programs currently collect and disseminate such information at various levels. Both domestically and internationally, the committee did not find a more comprehensive source for mineral information today than the Minerals Information Team of the USGS. In view of the need to examine and determine mineral criticality within the framework of the complete mineral life cycle, the committee also suggests several additional types of mineral information that could be collected by the federal government and some research avenues that would be helpful to pursue in support of this type of data collection.

The committee found further:

- Fully federally supported mineral information collection and research is needed by the public and private sectors to ensure unbiased and thorough reporting of available data.
- Accurate and timely mineral information with regard to factors affecting mineral availability in the short, medium, and long term is important to guide policy.
- Mineral information collection and research in the federal government is less robust than it could be, in part because the minerals data collection program of the Minerals Information Team does not have the status of a principal statistical agency.
- Many similarities exist between the function and form of the EIA and the Minerals Information Team, with the exception of the status of the EIA as a principal statistical agency with survey enforcement authority and organizational autonomy.
- Many international, private, nongovernmental, and federal government organizations, associations, and agencies supply various types of mineral information to the public, some for a fee. In the United States, the committee found unanimous agreement from private, academic, and federal professionals that the USGS Minerals Information Team is the premier source of mineral information, but that the quantity and depth of data have been reduced in recent years.

- Full information on the mineral life cycle, and the critical mineral cycle particularly, requires information on recycling and scrap generation and inventories of old scrap; in-use stocks; reserves and resources; downstream uses; subeconomic resources; material flows; and international information in each of these areas. Federal mineral information collection presently does not include these factors.
- Support of critical mineral information collection includes enhanced research in such areas as theoretical geochemical research, extraction and processing technology, remanufacturing and recycling technology, and stocks and flows of materials.
- Formal coordination and data exchange between federal agencies collecting mineral information is an objective to reduce redundancy and ensure transparency but is not currently fully effective.
- A potential model for the collection of mineral information is described in the ICAF report, which suggests a mineral "life-cycle" approach that offers specific challenges to the mineral sector and requires flexible federal approaches to mineral policy at various stages of the cycle, based on consistent mineral data collection.
- With respect to the professional pipeline of training to carry out the data gathering, analysis, research, and exploration needed to evaluate minerals and their criticality, industry, the government, and educational institutions face an existing and growing shortage of resource professionals entering the system.

REFERENCES

Anderson, M., and W. Seltzer, 2005. Federal statistical confidentiality and business data: Twentieth century challenges and continuing issues. Paper prepared for the Annual Meeting of the Population Association of America, New York, March 30.

Cammarota, D., 2007. Presentation to the Committee on Critical Mineral Impacts on the U.S. Economy. Washington, D.C., March 8.

DeYoung, J. 2007. Presentation to the Committee on Critical Mineral Impacts on the U.S. Economy. Washington, D.C., March 8.

DUSD(IP) (Deputy Undersecretary of Defense for Industrial Policy), 2003. Transforming the Defense Industrial Base: A Roadmap. Available online at *http://www.acq.osd.mil/ip/docs/transforming_the_defense_ind_base-report_only.pdf* (accessed May 22, 2006).

The Economist, 2006. Special Report: The ageing workforce. February 18:65-67.

EIA (Energy Information Administration), 2006. *Overview of the Energy Information Administration's National Energy Information System.* Prepared in support of "Challenges, Choices, Changes: An External Study of the Energy Information Administration," unpublished final report. Washington, D.C.

Ellis, M., 2007. Comment from the Industrial Minerals Association—North America, submitted to the committee, May 11. On file at the National Academies Public Access Records Office.

ICAF (Industrial College of the Armed Forces), 2006. AY 2005-2006 Industry Study Final Report: Strategic Materials. Washington D.C.: National Defense University. 30 pp. Available online at *http://www.ndu.edu/icaf/industry/reports/2006/pdf/2006_STRATEGIC_MATERIALS.pdf* (accessed June 27, 2007).

LeVier, M., 2007. Presentation to the Committee on Critical Mineral Impacts on the U.S. Economy (teleconference). Washington, D.C., March 8.

Murakami, S., 2007. Presentation to the Committee on Critical Mineral Impacts on the U.S. Economy. Washington, D.C., March 8.

NRC (National Research Council), 2003. *Future Challenges for the U.S. Geological Survey's Mineral Resources Program.* Washington, D.C.: The National Academies Press, 139 pp.

NRC, 2006. *Improving Business Statistics Through Interagency Data Sharing: Summary of a Workshop.* Washington, D.C.: The National Academies Press, 156 pp.

NRC, 2007. *Rising Above the Gathering Storm: Energizing and Employing America for a Brighter Economic Future.* Washington, D.C.: The National Academies Press, 592 pp.

NSF (National Science Foundation), 2005. Federal Scientists and Engineers 1998-2002: Detailed Statistical Tables. Available online at *http://www.nsf.gov/statistics/nsf05304/pdf/nsf05304.pdf* (accessed September 2007).

OMB (Office of Management and Budget), 2007a. Strengthening Federal Statistics. Available online at *http://www.whitehouse.gov/omb/inforeg/statpolicy/strengthening_fed_stats_fy2008.pdf* (accessed November 14, 2007).

OMB, 2007b. Statistical Programs of the United States Government Fiscal Year 2007. Washington, D.C.: Executive Office of the President. Available online at http://www.whitehouse.gov/omb/inforeg/07statprog.pdf (accessed July 18, 2007).

Shanks, W., 1983. *Cameron Volume on Unconventional Mineral Deposits.* Society of Economic Geologists, Society of Mining Engineers, American Institute of Mining, Metallurgical, and Petroleum Engineers, 246 pp.

U.S. Congress, House, 1974. Federal Energy Administration (FEA) Act. H.R. 11793. 93rd Congress. Public Law 93-275. Washington, D.C.: U.S. Government Printing Office.

U.S. Congress, Senate, 1977. Department of Energy Organization Act. S.Doc. 826. 95th Congress. Public Law 95-91. Washington, D.C.: U.S. Government Printing Office.

U.S. Congress, House, 1980. U.S. Minerals Vulnerability: National Policy Implications, A Report of the Subcommittee on Mines and Mining of the Committee on Interior and Insular Affairs. 96th Congress, 2nd Session.

USGS (U.S. Geological Survey), 2007. Mineral Information Team Information Handout. Available online at *http://minerals.usgs.gov/minerals/pubs/mit/mit.pdf* (accessed July 3, 2007).

CHAPTER 6

Conclusions and Recommendations

Minerals, or more specifically the mineral products derived from them, are essential to the functioning of modern processes and products. Some minerals are more essential than others, in the sense that they have few if any substitutes capable of providing similar functionality at similar costs.

The availability of these minerals is a function of geologic, technical, environmental and social, political, and economic factors. Some minerals are more prone than others to disruptive restrictions in supply.

It is this combination of importance in use and supply risk, and specifically the potential that an important mineral may be subject to supply restrictions, that motivated this study. The committee was charged to carry out a number of specific tasks identified in Chapter 1:

1. Identify the critical minerals and mineral products that are essential for industry and emerging technologies in the domestic economy.
2. Assess the trends in the sources and production status of these critical minerals and mineral products worldwide.
3. Examine actual or potential constraints, including but not limited to geologic, technologic, economic, and political issues, on the availability of these minerals and mineral products for domestic applications.
4. Identify the impacts of disruptions in supply of critical minerals and mineral products on the domestic workforce and economy.

5. Describe and evaluate the current mineral and mineral product databases and other sources of information available for decision making on mineral policy issues.
6. Identify types of information and possible research initiatives that will enhance understanding of critical minerals and mineral products in a global context.

Chapters 2 through 5 have examined the various dimensions of the overall task, and each chapter concluded with principal findings. This chapter presents the committee's principal conclusions, drawing on each previous chapter's findings, and summarizes the committee's recommendations following from these conclusions. Throughout its examination of these issues, the committee found it essential to consider minerals, and critical minerals, in the context of a global mineral and material cycle—from mineral ores at the mine to metallic and nonmetallic minerals in potentially recyclable materials and products.

The committee established parameters regarding a mineral's *importance in use* and *availability* (supply risk) to apply the criticality matrix to 11 minerals or mineral groups: copper, gallium, indium, lithium, manganese, niobium, platinum group metals (PGMs), rare earth elements (REs), tantalum, titanium, and vanadium. The committee did not have the time or resources to evaluate all potentially critical minerals. Instead, the committee selected the minerals identified above on the basis of two considerations. First, the set of minerals the committee examined had to illustrate the range of circumstances that the matrix methodology accommodates and considers. For example, in its selection of the minerals examined in this report, the committee considered minerals used in large quantities throughout the economy in traditional applications and others used in limited quantities in a small number of (often emerging) applications, minerals produced largely as by-products, and other minerals for which recycling of scrap is an important source of supply. Second, the set of minerals had to consist of those that, in the professional judgment of committee members, would likely be included in a more comprehensive assessment of all potentially critical minerals. The committee used a combination of quantitative measures and expert (qualitative) judgment in implementing the matrix methodology.

CONCLUSIONS

Defining Criticality

The committee concludes that **all minerals and mineral products could be or could become critical to some degree, depending on their importance and availability**—in the sense that the chemical and physical properties they provide are essential to a specific product or use or more broadly, that specific minerals are an essential input for a national priority (for example, national defense) or for an industry, or may be important (or have the potential to become important) to a region or the nation as a whole. Materials derived from minerals are essential to the performance of nearly all products and services we take for granted—cellular telephones, automobiles, home appliances, computers and other electronic products, and aircraft, for example. The degree of a mineral's importance can vary considerably over time as technologies and the economy evolve and change.

The committee also concludes, however, that more useful **from the federal perspective is the concept of a critical mineral as one that is both essential in use *and* subject to supply restriction**. In other words, the key determinants of criticality here are importance in use and availability. Based on these determinants, the committee developed a methodology—a 'criticality matrix'—for assessing the criticality of specific minerals and identified the information requirements for implementing this methodology. The matrix has two dimensions. The first (vertical axis) represents the degree of importance of a mineral or, equivalently, the impact of a supply restriction. The second dimension (horizontal axis) represents the degree of supply risk or the risk of a supply restriction.

This methodology emphasizes that criticality is a relative concept in that minerals are more or less critical, rather than critical or not critical. At any time, and for any organization or nation, some minerals will be more critical than others. Over time, **the criticality of a specific mineral**

can and likely will change as production technologies evolve and new products are developed.

Furthermore, the committee concludes that in implementing the methodology to assess criticality, **it is important to distinguish among three time or adjustment periods**. In the *short term* (period of a few months to a few years), mineral markets and in turn prices are influenced primarily by unexpected changes in mineral demand, such as the largely unanticipated increase in Chinese mineral demand over the last several years, and by unexpected shortfalls in production due to technical or other problems at existing mines and production facilities. In the short term, from the perspective of a mineral market as a whole, mineral users and producers are constrained by their existing production capacity, and therefore, unexpected changes in demand or supply are reflected largely in inventories held by producers, users, and commodity exchanges.

In the *medium term* (a few years, but no more than about a decade), markets respond to short-term developments but still in a relatively limited manner; for example, if a mineral's availability has become restricted, mineral users make any easy substitution for this mineral, and mineral producers bring into production any easy-to-develop, higher-cost sources of the restricted mineral (e.g., higher-cost scrap that previously was not recycled; and higher-cost, known but underdeveloped mineral deposits). In the medium term, mineral users and producers are essentially limited by existing technologies and known primary and secondary mineral resources.

Over the *long term* (roughly a decade or more), mineral users and producers can respond more significantly to changes in mineral availability through conscious decisions about whether and to what degree to invest in innovative activities in mineral exploration, mine development, mineral processing, product design and manufacturing, and recycling technology and policy.

Understanding Importance in Use or the Impact of a Supply Restriction

Users demand minerals and mineral attributes for the functionality they provide—their chemical and physical properties in specific applications such as strength, corrosion resistance, electrical conductivity, low density, and so on. As noted at the beginning of this chapter, some minerals are more essential than others in the sense that they have few if any substitutes capable of providing similar functionality at similar costs. **The greater the difficulty, expense, or time it takes for material substitution to occur, the more critical a mineral is to a specific application or product—or analogously, the greater is the impact of a supply restriction.**

The impact of a specific supply restriction, in other words, depends on the nature of the restriction. A supply restriction can occur in two general forms. First, demand can increase and outstrip existing production capacity (a demand shock). Second, in what normally would be considered a disruption, a material that previously was available becomes unavailable (a supply shock). In either case, it is possible that a mineral or mineral product becomes physically unavailable; in this situation, the product a user makes cannot be manufactured, sold, and then used by the prospective purchaser. More typically, however, a mineral or mineral product remains physically available, but at a higher price. In this situation, supply will be reallocated to those users willing to pay more for a mineral or mineral product and away from lower-valued uses.

The specific impact of a supply restriction will depend on circumstances: Is the mineral physically unavailable, or have prices increased? If prices rise, by how much? How flexible or inflexible is demand (that is, how easy or difficult is it to substitute for the restricted mineral)? Finally, time is important. In the short term, mineral users will be relatively limited in the degree to which they can adjust to physical unavailability or higher prices for a mineral or mineral product. Users are constrained by the flexibility of their production processes that use minerals as inputs. Most production processes are relatively inflexible in the short term. A facility that manufactures aluminum cans, for example, cannot immediately reduce the amount of aluminum it uses per can or convert itself into glass bottle making facility. In the medium term, users have somewhat more flexibility. An aluminum can-making facility might be able to invest in existing technology that uses less aluminum per can than its facility currently requires. Alternatively, it might decide to become a glass bottle-making facility. Over the long term, users of minerals and mineral products will be relatively most flexible to respond to a supply restriction. There is time for a facility that manufactures aluminum cans to innovate and develop a process for using less aluminum per can than previously.

In any of these adjustment periods, the types of possible effects include impacts on:

- *Domestic production of minerals and mineral products:* there may be opportunities for increased domestic production of the mineral or mineral product whose supply has been restricted (higher-cost but previously uneconomic primary or secondary production).
- *Domestic users of minerals or mineral products (typically producers of semifabricated products and manufacturers of final products):*
  - — Lost production due to lack of availability or higher costs (use will be concentrated in higher-valued uses of a mineral or mineral product);
  - — Higher costs of production, which producers may or may not be able to pass along to consumers;
  - — Slower growth than otherwise in emerging-use industries;
  - — Less employment than otherwise in industries using minerals and mineral products as inputs;
  - — Ultimately lower value added in those sectors using minerals and mineral products, and lower gross domestic product (GDP), although the impact on GDP of a supply disruption for any single mineral or mineral product will be small from the perspective of the national economy;
  - — Higher costs or reduced availability of products related to national defense.
- *Domestic purchasers of goods containing minerals and mineral products:* there may be fewer purchases or more expensive purchases because goods have become more expensive (in either case, purchasers are worse off than previously).

The committee did not attempt to quantify these effects. To do so would have required detailed and separate economic impact analyses for each specific circumstance, and the committee was not constituted with sufficient expertise to carry out this type of quantitative analysis. However, the committee notes that the largest impacts on national employment and GDP would come from supply restrictions on minerals and mineral products used in large quantities; of the minerals the committee examined using its criticality methodology, copper falls into this category, even though copper did not qualify as critical in the committee's eyes because its supply risk is low. Other minerals that the committee believes would be evaluated similarly include iron ore, aluminum, and aggregates.

Understanding Availability and Supply Risk

Fundamentally, minerals are a primary resource in that we obtain them from the Earth's crust. At any point in time, however, minerals—or more precisely the mineral products obtained from them—are available as secondary resources through recycling of obsolete or discarded products and materials. Finally, from the perspective of a nation, mineral products are available as tertiary resources embodied in imported products or imported scrap. The U.S. economy obtains minerals and mineral products in all three forms—primary, secondary, and tertiary. Although the United States has been and remains an important producer of primary and secondary minerals, it also relies on imports for a number of primary and tertiary minerals.

For primary production worldwide and in the United States, mineral exploration, mining, and mineral processing are sectors whose fortunes change significantly from year to year because of the strong link between mineral demand and economic growth. In periods of especially strong economic growth, mineral use in general expands more quickly than production capacity, tending to drive up mineral prices, whereas in periods of slower growth or recession, mineral use tends to grow more slowly than production capacity and prices tend to fall. Given the fragility of the balance between demand and supply, mineral prices tend to swing significantly from one year to another. Since early in this decade, the mineral sector overall has experienced an extended boom (and relatively high mineral prices) due to a number of factors, including unexpectedly large increases in mineral demand in China and some other countries and unexpected interruptions in production at a number of mines due to technical problems and other factors.

The level and location of mine production today depend on the level and location of mineral exploration in the past. The level of exploration tends to follow changes in mineral prices, but usually with a short time lag. The composition of exploration activity varies with mineral prices. In recent years during a period of relatively high mineral prices, exploration by small exploration companies (termed "juniors") in riskier and more remote locations has increased proportionately more than exploration by larger and more established mining companies. Conversely, when mineral prices fall, exploration by junior companies tends to fall proportionately more than that by larger companies, resulting in relatively less exploration in remote locations and more exploration in proximity to existing mines. The geographic location of exploration and mining also evolves over time. In recent years, relatively more exploration and mining has occurred outside the established areas of Australia, Canada, and the United States.

Turning from primary to secondary production, recycling tends to be concentrated close to semifabrication and metal manufacturing facilities and close to urban centers to take advantage of the creation of scrap when buildings are demolished and products are discarded. As a result, most metal recycling occurs in industrialized economies where the majority of metal use historically has occurred. Nevertheless, a significant amount of recycling occurs in developing economies, where perhaps a larger percentage of the available scrap is actually recycled than in industrialized economies. Given the long-term trend of increasing mineral use and low rates of recycling, recycled materials cannot presently meet a large proportion of demand for most materials. Over time, as products used in developing economies become available for recycling, we can expect scrap flows to increase and the location of recycling to become more geographically diverse than at present.

In considering supply risk and implementing the matrix methodology, as noted above, the committee found it essential to distinguish between short- and medium-term availability of minerals and mineral products, on the one hand, and long-term availability, on the other. **In the *short and medium term*, there may be significant restrictions to supply for at least five reasons.** First, **demand may increase significantly**, and if production already is occurring at close to capacity, then either a mineral may become physically unavailable or, more likely, its price will rise significantly—demand can increase more quickly than production capacity can respond. Second, an increase in demand due to growth in new applications of a mineral may be especially restrictive or disruptive if preexisting uses were small relative to the new use (**thin markets**). Third, supply may be prone to restriction if **production is concentrated**; if concentrated in a small number of mines, supply may be prone to restriction if unexpected technical or labor problems occur at a mine; if concentrated in the hands of a small number of companies, supply may be prone to restriction by opportunistic behavior of companies with market power; if concentrated in the hands of a small number of producing countries, supply may be prone to restriction due to political decisions in the producing country. Fourth, if **mine production comes predominantly in the form of by-product production**, then the output over the short term (and perhaps even longer) may be insensitive to changes in market conditions for the by-product because the output of a by-product is largely a function of market conditions for the main product. Finally, the **lack of available old scrap for recycling or of the infrastructure required for recycling** makes a market more prone to supply restriction than otherwise.

An additional factor, import dependence, often is cited as an indicator of vulnerable supply and has carried the implication that imported supply may be less secure than domestic supply. The committee concludes that **import dependence by itself is not a useful indicator of supply risk.** In fact, import reliance may be good for the U.S. economy if an imported mineral has a lower cost than the domestic alternative. Rather, for imports to be vulnerable to supply restriction, some other factor must be present that makes them vulnerable to disruption—for example, supply is concentrated in one or a small number of exporting nations with high political risk or in a nation with such significant growth in internal demand that formerly exported minerals may be redirected toward internal, domestic use. However, imports may be no less secure than domestic supply if they come from a diverse set of countries or firms or if they represent intracompany transfers within the vertical chain of a firm (for example, imported metal concentrate to be smelted and refined at a company's domestic processing facilities).

Over the *longer term*, the availability of minerals and mineral products is largely a function of investment and the various factors that influence the level of investment and its geographic allocation and success. An important investment is that in education and research, and the committee suggests that the **long-term availability of minerals and mineral products also requires continued investment in mineral education and research.**

Education and research contribute to determining long-term mineral availability for both primary and secondary resources in all of their dimensions. For primary resources, the first important dimension is geologic availability (in what quantities, concentrations, and mineralogical forms does a mineral exist in Earth's crust?). Education and research of course do not determine whether and in what form a mineral occurs in Earth's crust; rather education and research determine our *knowledge* of Earth's crust. The second determinant is technical availability (does the technology exist to extract and process the element or mineral?). Technical availability depends on investment in technological knowledge. The third determinant is environmental and social availability (can we mine and process minerals such that the consequences of these activities on local communities and on the natural environment are consistent with social preferences and requirements?). Environmental and social availability depends on investment in activities that appeal to social preferences and that develop means for carrying out mining and mineral processing in socially acceptable ways. The fourth determinant is political availability (to what extent do public policies influence mineral supply?). Political availability depends on investment in the design of public policy and on the political decisions governments make that influence the level and location of production. The fifth and final determinant is economic availability (can we produce minerals and mineral products at prices that users are willing and able to pay?). In some sense, economic availability reflects the combined effects of the other four determinants of availability.

For secondary resources over the longer term, availability depends on four of the same above factors. Technology in the secondary resources sector is far behind that in the primary sector, and many gains are to be had by investing additional engineering time and effort. On the environmental and social front, recycling needs to occur with a greater degree of urgency, and making changes in this area is largely a social challenge. Politically, attention needs to be paid to understanding the national implications of resource scarcity, to providing the funds to better characterize the secondary resource, and to better evaluate opportunities for domestic recovery of secondary materials. Finally, it will be necessary to create economic incentives to make better use of the secondary resources now above the ground and in use, but often more costly to use at present than imported virgin material. Well-designed and competently directed research into improved recycling technologies may prove an effective tool in the reduction of our dependence on imports of critical minerals.

Implementing the Mineral Criticality Matrix

The committee applied its criticality matrix methodology to 11 minerals or mineral families it considered candidates for criticality. The committee acknowledges the existence of numerous other minerals that individuals, industrial sectors, organizations, or government officials might consider critical to their particular needs or requirements now or in the future. At a practical level, the committee did not have the resources for comprehensive analysis of all minerals using its methodology.

In evaluating these minerals or mineral families, the committee took a short- and a medium-term perspective—that is, within the next decade, what are the risks of a supply restriction, and how significant would the impact of restrictions be should they occur? Of the 11 minerals or mineral families the committee examined, **those that exhibit the highest degree**

of criticality at present are: indium, manganese, niobium, PGMs, and REs. The committee studied PGMs and REs in some depth, while it examined indium, manganese, and niobium in a more limited manner. Each of these minerals has a slightly different story in terms of importance in use (impact of a supply restriction) and availability (supply risk), the two dimensions of criticality.

PGMs—consisting primarily of platinum, palladium, and rhodium—are essential in automotive catalysts. Palladium can partially substitute for platinum in gasoline vehicles. Palladium cannot be substituted for platinum in diesel vehicles. Rhodium has no known substitutes in the control of NO_x emissions. PGMs also are essential determinants of product quality in several industrial applications (the production of fertilizers, explosives, and petrochemicals). PGMs are mined almost exclusively in South Africa and Russia, and are typically mined as coproducts. The United States has two small PGM mines and a minor quantity of subeconomic PGM resources. Recycling occurs, primarily of spent automotive catalysts, but this amount is modest relative to annual use. The PGM market is relatively small, with annual worldwide mine production on the order of 200,000 kilograms.

REs are essential, with few if any good substitutes, in automotive catalytic converters, permanent magnets, and phosphors used in medical imaging devices, televisions, and computer monitors. The RE market is fragile because it is small—worldwide mine production in 2006 was on the order of 100,000 metric tons. U.S. manufacturers import REs predominantly from China. Very little recycling occurs. The United States has significant RE resources, but at present these resources are subeconomic.

Indium has no adequate substitutes for flat-panel displays. This use has experienced rapid growth in recent years. Worldwide mine production is small—some 500 metric tons in 2006, largely as a by-product of zinc mining and processing. The indium that U.S. manufacturers use comes primarily from China, Canada, Japan, and Russia. Very little indium is recovered through recycling.

Manganese has no satisfactory substitutes as a hardening element in various types of steel. It is not mined at present in the United States. The majority of U.S. imports comes in the form of ore from Gabon and South Africa and ferromanganese from South Africa, China, Brazil, and France. U.S. manganese resources are subeconomic. Some manganese is recovered as a part of ferrous and nonferrous scrap recovery; almost none of this recovery is for manganese in particular but rather for the steel or other nonferrous metal of which manganese is a minor element.

Niobium is used in carbon, high-strength low-alloy (HSLA), and stainless steels. It also is used in superalloys for aircraft engines. Where substitution is technically possible, performance is sacrificed. Niobium use in HSLA steels has fallen considerably, but has increased in superalloys. Niobium is not mined in the United States, at least not in any significant quantity. U.S. users import the majority of their niobium from Brazil and to a lesser extent from Canada. The niobium market is small; estimated 2006 mine production was on the order of 60,000 metric tons. Known U.S. resources are very small and subeconomic. Significant recycling of niobium from niobium-containing steels and superalloys occurs; very little of this recycling is targeted at niobium in particular but rather for the steel or superalloy itself.

On the basis of these applications of the methodology, the committee concludes that **the criticality matrix methodology is a useful conceptual framework for evaluating a mineral's criticality in a balanced manner in a variety of circumstances that will be useful for decision makers in the public and private sectors**. Decision makers should be prepared to reevaluate a mineral's criticality whenever one of the underlying determinants of criticality changes or appears likely to change. In the short to medium term, the most likely factors to change are, first of all, demand, which could increase sharply if a new application is developed for a specific mineral and, second, the degree to which a mineral's production is concentrated in a small number of companies or countries, which in turn might be prone to opportunistic behavior. A more nuanced and quantitative version of the matrix could be established and used as part of the federal program for mineral data collection, analysis, and dissemination.

Assessing Information and Research Needs

In the progress of this study, the committee has frequently compared the constrained scope and depth of information on minerals with the broad scope and great depth of financial information acquired and analyzed by the federal government. The usefulness of this financial information by governments, industries, and many other users suggests that an enhanced information program on minerals could be more broadly and deeply beneficial as well. The mineral information available at present is used widely but is also acknowledged to be considerably less detailed than is desirable. This is particularly the case for mineral information related to other countries, where high-quality data are essential for accurate determinations of criticality for U.S. industries and for the country as a whole.

A large number of government and nongovernmental, international, and domestic organizations collect and disseminate information and databases relevant for decision making on critical minerals and other mineral policy issues for public and private use. The consensus view of private, academic, and federal professionals is that the U.S. Geological Survey (USGS) Minerals Information Team is the most comprehensive, responsible, and responsive source of mineral information internationally, but that the quantity and depth of its data and analysis have fallen in recent years, due at least in part to reduced or static budgets and associated reductions in staff and data coverage.

In its evaluation of information and research needs, the committee concludes the following:

- **Decision makers in both the public and the private sectors need continuous, unbiased, and thorough mineral information provided through a federally funded system of information collection and dissemination.**
- **The effectiveness of a government agency or program is dependent on the agency's or program's autonomy, its level of resources, and its authority to enforce data collection.** In the committee's view, **federal information gathering for minerals at**

present does not have sufficient authority and autonomy to appropriately carry out data collection, dissemination, and analysis. In particular, the committee concludes that USGS Minerals Information Team activities are less robust than they might be, in part because it does not have status as a "principal" statistical agency.

- **More complete information needs to be collected, and more research needs to be conducted, on the full mineral life cycle.** The committee includes its specific recommendations in the following section. A common theme in these recommendations is the value of an investment in material flow accounting to better quantify stocks, flows, and uncertainty for primary, secondary, and tertiary resources.

RECOMMENDATIONS

Recognizing the dynamic nature of mineral supply and demand and of criticality, and in light of the conclusions above, the committee makes the following recommendations:

1. The federal government should enhance the types of data and information it collects, disseminates, and analyzes on minerals and mineral products, especially as these data and information relate to minerals and mineral products that are or may become critical.

In particular, more attention than at present needs to be given to those areas of the mineral life cycle that are underrepresented in current information-gathering activities, including: reserves and subeconomic resources; by-product and coproduct primary production; stocks and flows of secondary material available for recycling; in-use stocks; material flows; international trade, especially of metals and mineral products embodied in imported and exported products; and related information deemed appropriate and necessary. Enhanced mineral analysis should include periodic assessment of mineral criticality over a wider range of minerals and in greater depth than was possible for this committee to undertake, using the committee's methodology or some other suitable method.

2. The federal government should continue to carry out the necessary function of collecting, disseminating, and analyzing mineral data and information. The USGS Minerals Information Team, or whatever federal unit might later be assigned these responsibilities, should have greater authority and autonomy than at present. It also should have sufficient resources to carry out its mandate, which would be broader than the Minerals Information Team's current mandate if the committee's recommendations are adopted. It should establish formal mechanisms for communicating with users, government and nongovernmental organizations or institutes, and the private sector on the types and quality of data and information it collects, disseminates, and analyzes. It should be organized to have the flexibility to collect, disseminate, and analyze additional, nonbasic data and information, in consultation with users, as specific minerals and mineral products become relatively more critical over time (and vice versa).

The Energy Information Administration provides a potential model for such an agency or administrative unit. The federal government should consider whether a comparable mineral information administration would have status as a principal statistical agency and, if not, what other procedures should be investigated and implemented to give an agency with the mandate to collect mineral data and information greater autonomy and authority, as well as sufficient resources to carry out its mandate. In the globalized mineral market, it is essential that the United States has a central authority through which to conduct outreach and exchange programs on mineral data with international counterparts and to collect and harmonize data from international sources. Combined U.S. government and foreign government efforts are likely to provide the most accurate, uniform, and complete data sets of this information over time and thereby provide adequate information to all communities concerned about future global mineral or material supply and demand trends.

3. Federal agencies, including the National Science Foundation, Department of the Interior (including the USGS), Department of Defense, Department of Energy, and Department of Commerce, should develop and fund activities, including basic science and policy research, to encourage U.S. innovation in the areas of critical minerals and materials and to enhance understanding of global mineral availability and use.

Without renewed federal commitment to innovative mineral research and education, it is doubtful whether the recommended activities regarding mineral information will be sufficient for the nation to successfully anticipate and respond to possible short- to long-term restrictions in mineral markets.

The committee recommends the following additional initiatives in this regard:

- Funded support for scientific, technical, and social scientific research focusing on the entire mineral life cycle, especially those specific areas identified in Recommendation 1; and
- Cooperative programs involving academic organizations, industry, and government to enhance education and applied research.

Appendixes

APPENDIX A

Biographical Sketches of Committee Members and Staff

COMMITTEE MEMBERS

RODERICK G. EGGERT (*Chair*) is professor and director of the Division of Economics and Business at the Colorado School of Mines, where he has taught since 1986. He was editor of *Resources Policy*, an international journal of mineral economics and policy, from 1989 to 2006. Previously he taught at The Pennsylvania State University and held research appointments at Resources for the Future (Washington, D.C.) and the International Institute for Applied Systems Analysis (Austria). He has a B.A. in earth sciences from Dartmouth College, an M.S. in geochemistry and mineralogy from Pennsylvania State University, and a Ph.D. in mineral economics from Penn State. His research and teaching have focused on various aspects of mineral economics and public policy, including the economics of mineral exploration, mineral demand, mining and the environment, microeconomics of mineral markets, and most recently mining and sustainable development. He served for two terms on the Committee on Earth Resources of the National Research Council.

ANN S. CARPENTER was hired as president and chief operating officer for U.S. Gold Corporation in October 2005. From 2003 until 2005, Ms. Carpenter was an independent consultant in the mining industry, focusing on resource assessment, evaluations, and project development for properties in the United States, Mexico, and South America. From November 1997 to 2003, she was the vice president of exploration and business development for NCGI, a private mining company. With more than 26 years of experience in mineral development activities worldwide, her work focuses on resource calculations; engineering evaluations and studies; permitting requirements; and regulatory, legislative, and policy evaluations regarding mineral development here in the United States and overseas. She is president of the Northwest Mining Association; past president and current adviser of the Women's Mining Coalition; and foundation chairman and past officer of the Geological Society of Nevada. Ann is also an active member of the Society for Mining, Metallurgy, and Exploration (SME), Mining and Metallurgical Society of America (MMSA), and Prospectors and Developers Association of Canada (PDAC). She has testified before the U.S. Congress on matters regarding domestic mineral resource potential and development, permitting, mineral policy, and sustainable development. She earned her B.S. in geology from Montana State University in 1980, and completed advanced geologic studies at Mackay School of Mines in Reno, Nevada, from 1981 to 1983.

STEPHEN W. FREIMAN left the National Institute of Standards and Technology (NIST) in 2006 to start a consulting business (Freiman Consulting Inc.). He had been at NIST since 1978 (at that time called the National Bureau of Standards) where he worked primarily on the fracture of brittle materials. From 1992 to 2002, Dr. Freiman served as Chief of the Ceramics Division at NIST, and from 2002 to 2006 he served as deputy director of the Materials Science and Engineering Laboratory (MSEL). Dr. Freiman has published more than 150 papers focusing on the mechanical properties of brittle materials. He was the first chairman of the ASTM Subcommittee addressing brittle fracture and a past chair of the Versailles Project on Advanced Materials and Standards (VAMAS) steering committee. He is a fellow and a past president of the American Ceramic Society. Dr. Freiman graduated from the Georgia Institute of Technology with a

B.ChE. and an M.S. in metallurgy. After receiving a Ph.D. in materials science and engineering from the University of Florida in 1968, he worked at the IIT Research Institute and the Naval Research Laboratory.

THOMAS E. GRAEDEL (NAE) has been a professor of industrial ecology, chemical engineering, and geophysics at Yale University since 1997. From 1969 to 1984, Dr. Graedel was a member of the technical staff, and from 1984 to 1996, a distinguished member of the technical staff at AT&T Bell Laboratories. Dr. Graedel's research interests have included solar physics; chemical kinetic modeling of gases and droplets in Earth's atmosphere; corrosion of materials by atmospheric species; atmospheric change; and industrial ecology and sustainability science. He and his colleagues have assessed regional and global cycles for metals including copper, chromium, zinc, lead, and silver, determining the stocks available in different types of reservoirs and the related flows. Ongoing work treats a number of other metals, including iron, nickel, stainless steel, and tungsten. He is the author or coauthor of 13 books and more than 300 technical papers in various scientific journals. Dr. Graedel received his B.S. (chemical engineering) from Washington State University in 1960, his M.A. (physics) from Kent State University in 1964, and his M.S. and Ph.D. (astronomy) from the University of Michigan in 1967 and 1969, respectively.

TERENCE P. MCNULTY (NAE) has been president of T.P. McNulty and Associates, Inc., since 1989. His company's work for a global client base includes process engineering in base and precious metals, uranium, nonmetallic minerals, and industrial chemicals; direction of research programs; management consulting and strategic planning; project management; plant audits; and assistance in commercializing new technologies. From 1983 to 1988, he served as president and chief executive officer of Hazen Research, Inc., and from 1980 to 1983 he was vice president of technical operations for Kerr-McGee Chemical Corp. From 1960 to 1980 he was employed by the Anaconda Company. Dr. McNulty has two patents in copper metallurgy and 40 publications in the fields of (1) minerals processing and the extractive metallurgy of iron, copper, uranium, and precious metals; (2) process control; (3) energy conservation; (4) mineral industry trends; (5) waste treatment; (6) project management; and (7) technology development. He received his B.S. in chemical engineering in 1960 from Stanford University, his M.S. in metallurgical engineering in 1963 from the Montana School of Mines, and his D.Sc. in extractive metallurgy in 1966 from the Colorado School of Mines.

DREW A. MEYER is the former vice president, Marketing & Transportation Services, Vulcan Materials Company, from which he retired in March 2007. At Vulcan, he was responsible for the Marketing, Market Research, Marketing Support Services, Transportation Sales, and Support Services Departments, and he led economic forecasting and analysis for the Construction Materials Group. Mr. Meyer has spent more than 38 years with Vulcan Materials Company during which time he has worked at the corporate, group, and division levels both domestically and overseas. Mr. Meyer has been an active participant in serving the construction aggregates industry through its associations and was elected to honorary life membership in the National Stone, Sand & Gravel Association (NSSGA) in January 2004. Prior to its merger with NSSGA, he served in a number of leadership positions in the National Aggregates Association including treasurer and vice chairman. In 2003, *Aggregate Manager Magazine* selected Drew as the AGGMAN Professional of the Year for 2002. He is a past chairman of SME's Construction Materials and Aggregates Committee, a recipient of the President's Citation for Outstanding Leadership and Service to SME in 2003, a member of the SME Board of Directors, and vice president of the SME Foundation. In addition to SME, Mr. Meyer is a member of the American Marketing Association and the National Association of Business Economists; he is on the Board of Directors of the Minerals Information Institute. He has authored and coauthored articles and made numerous presentations on subjects related to the stone industry and to the extraction, processing, and consumption of magnetic metals from municipal solid waste. He is a graduate of The Pennsylvania State University with B.S. and M.S. degrees in mineral economics.

BRIJ M. MOUDGIL (NAE) is a distinguished professor and alumni professor of materials science and engineering, and director of the Particle Engineering Research Center at the University of Florida (UF), Gainesville. Dr. Moudgil also serves as the director of the UF Mineral Resources Research Center. Prior to joining the University of Florida faculty in 1981, Dr. Moudgil was associated with the Occidental Research Corporation in California as a research engineer. He was a visiting research fellow at E.I. DuPont de Nemours & Co. in Wilmington, Delaware, in 1988-1989. His current research interests include nanoparticulate processing and separation technology for enhanced performance in mineral, chemical, microelectronics, pharmaceutics, advanced materials, and resource recovery and waste disposal applications. Dr. Moudgil is a member of several professional societies (AIChE, SME, ACS, ACerS, MRS, TAPPI). He is currently serving as president of the SME. He has published more than 200 technical papers, has been awarded 14 patents, and has edited 9 books. He received his undergraduate training in metallurgy at the Indian Institute of Science, Bangalore. He continued his graduate studies at Columbia University, New York, and received M.S. and Eng.Sc.D. degrees in mineral engineering-interfacial phenomena applied to particulate processing.

MARY M. POULTON is professor and department head at the Department of Mining and Geological Engineering, University of Arizona, where she has been a faculty member since 1990. Her research interests include neural networks, geosensing and geophysics, mineral exploration and production, resource characterization, and water and energy resources exploration. She has published numerous papers in refereed journals and books, and has edited a book on geophysical data processing. In addition to her service at the university, Dr. Poulton has worked in consulting capacity for Neural Optimization Applied Hydrology, LLC, in the area of water and energy optimization, environmental impacts, and water security. She received her B.S. (1984), M.S. (1987), and Ph.D. (1989) in geological engineering from the University of Arizona.

LEONARD J. SURGES has been director general, Industry Analysis and Business Development, at Natural Resources Canada since 2005 where he is responsible for a policy branch within the Minerals and Metals Sector, including commodity and market analysis for metals and nonmetallic minerals, and international cooperation, trade, and investment relations. From 2002 to 2005, he was director of sustainable development and product policy at Noranda Inc./Falconbridge Limited; he was employed from 1992 to 2002 with Noranda Inc. with responsibility for environmental policy and management. An assignment from 1990 to 1992 with Environment Canada as a mining and smelting senior engineer was preceded by 11 years with Brunswick Mining and Smelting Corporation Limited. Membership in professional societies includes the Association of Professional Engineers of New Brunswick; the Canadian Institute of Mining, Metallurgy and Petroleum, and The Minerals, Metals and Materials Society. His professional career has overlapped with military service in the Canadian Forces where he most recently served with the Royal New Brunswick Regiment (North Shore) and held appointments as company commander and deputy commanding officer with the rank of major. He received his bachelor of applied science (B.A.Sc.) degree in metallurgical engineering from the University of British Columbia in 1979.

NRC STAFF

ELIZABETH A. EIDE, senior program officer, is a geologist with specialization in isotope geochronology applied to crustal processes, including oil and gas exploration. Prior to joining the National Research Council, she worked for 12 years at the Geological Survey of Norway where she constructed and managed an $^{40}Ar/^{39}Ar$ geochronology laboratory and managed personnel, budget, and research matters as team leader-researcher for two departments. She received her Ph.D. in geology from Stanford University and a B.A. in geology from Franklin and Marshall College.

NICHOLAS D. ROGERS is a research associate with the Board on Earth Sciences and Resources, National Research Council. He received a

B.A. in history, with a focus on the history of science and early American history, from Western Connecticut State University in 2004. He began working for the National Academies in 2006 and has primarily supported the Board on Earth Sciences and Resources on earth resource issues.

APPENDIX B

Workshop Agenda and Participants

AGENDA

Committee on Critical Mineral Impacts on the U.S. Economy

Board on Earth Sciences and Resources
National Research Council of the National Academies
National Academy of Sciences Building, Members Room
2100 C Street, NW, Washington, D.C.
March 7-9, 2007

Day 1—Wednesday, March 7, 2007

08:00-09:00 **CLOSED SESSION (Committee and NRC Staff only, working breakfast)**

09:00-16:30 **OPEN SESSION (Open to public)**

09:00-09:15 Welcome and introductions *Rod Eggert, Chair*

09:15-12:15 **Critical mineral sources and materials flow**—Panel 1
Moderated by Tom Graedel
Presentations and discussion (25-minute presentations)

Jeremiah Johnson (Hawaii Island Sustainable Energy Initiative)

	Shinsuke Murakami (National Institute for Environmental Studies, Japan) Mark Barton (University of Arizona) Larry Stevens (Consultant, Indium Corporation of America)	
12:15-13:00	*Lunch*	
13:00-16:00	**End users**—Panel 2 Presentations and discussion (25-minute presentations) Phil Jones (Imerys) Ivan Herring (General Motors Global Commodities) John Benner (National Renewable Energy Laboratory) James Marder (ASM International)	*Moderated by Stephen Freiman*
16:00-16:30	Concluding remarks	*Rod Eggert*

End of open session

16:30-20:00	**CLOSED SESSION (Committee and NRC Staff only)**	

Day 2—Thursday, March 8, 2007

08:00-09:00	**CLOSED SESSION (Committee and NRC Staff only)**	
09:00-16:30	**OPEN SESSION (Open to public)**	
09:00-09:15	Welcome and introductions	*Rod Eggert, Chair*
09:15-12:15	**Potential constraints on availability and adjustment mechanisms**—Panel 3 Presentations and discussion (25-minute presentations) Rich Heig (Rio Tinto) Glenn Miller (University of Nevada Reno)	*Moderated by Ann Carpenter*

	Lew Sloter (Department of Defense, Office of the Secretary of Defense) Marc LeVier (Newmont Mining Corporation)
12:15-13:00	*Lunch*
13:00-16:00	**Information, data, and research**—Panel 4 *Moderated by Leonard Surges* Presentations and discussion (25-minute presentations) John DeYoung and Dave Menzie (U.S. Geological Survey) David Cammarota (Department of Commerce) Jason Goulden (Metals Economics Group)
16:00-16:30	Concluding remarks — *Rod Eggert*

End of open session

16:30-17:30 CLOSED SESSION (Committee and NRC Staff only)

Day 3—Friday, March 9, 2007

08:00-14:30 CLOSED SESSION (Committee and NRC Staff)

OTHER PARTICIPANTS

Mark Barton	University of Arizona
John Benner	National Renewable Energy Laboratory
David Cammarota	U.S. Department of Commerce
Mike Canty	U.S. Department of Energy
Leslie Coleman	National Mining Association
John DeYoung	U.S. Geological Survey
John Dilles	Oregon State University
Mark Ellis	Industrial Minerals Association
Nora Foley	U.S. Geological Survey

Jason Goulden	Metals Economics Group
Linda Gundersen	U.S. Geological Survey
Rich Heig	Rio Tinto
Ivan Herring	GM Global Commodities
Ju Jin	Chinese Embassy
Jeremiah Johnson	Hawaii Island Sustainable Energy Initiative
Kate Johnson	U.S. Geological Survey
Phil Jones	Imerys
Mike Kaas	U.S. Bureau of Mines (retired)
Patrick Leahy	U.S. Geological Survey
Marc LeVier	Newmont Mining Corporation
James Marder	ASM International
David Menzie	U.S. Geological Survey
Glenn Miller	University of Nevada-Reno
Shinsuke Murakami	National Institute for Environmental Studies, Japan
Lauren Pagel	Earthworks
Carol Raulston	National Mining Association
David Russo	U.S. Geological Survey
Essie Schloss	Defense Logistics Agency
Lew Sloter	U.S. Department of Defense
Larry Stevens	Indium Corporation of America
Katie Sweeney	National Mining Association

APPENDIX C

Glossary

Alloy (metal, super)—A mixture of two or more chemical elements, at least one of which is a metal (Oxford Dictionary and Thesaurus, 1996).

Base metals—Any of the more common and more chemically active metals (e.g., lead, copper; Bates and Jackson, 1987).

By-product—Material of some economic value produced in a process that is focused on extracting another material. For example, indium is a by-product of zinc processing (available online at *http://www.platinum.matthey.com/production/1048863442.html*, accessed September 27, 2007).

Carbonatite deposit—A carbonate rock of apparent magmatic origin, generally associated with kimberlites and alkaline rocks (Bates and Jackson, 1987).

Catalytic converter—A device incorporated in the exhaust system of a motor vehicle, with a catalyst for converting pollutant gases into harmless products (Oxford Dictionary and Thesaurus, 1996).

Commodity—A physical substance such as grain, fuel, or minerals that is interchangeable with a product of the same type and that investors buy and sell, often through futures contracts (Bates and Jackson, 1987).

Conductivity—A measure of the ease with which a conduction current can be made to flow through a material under the influence of an applied electric field (Bates and Jackson, 1987).

Coproducts—Different minerals that occur in a single ore body that are extracted and/or processed together or sequentially with one another because of mineral or process similarities. Contrast to a mineral 'by-product' which is the residual of, or secondary to, extraction and production of a primary mineral of interest.

Critical minerals—Those that are both essential in use and subject to considerable supply risk.

Dielectric—A material in which displacement currents predominate over conduction currents (i.e., an insulator; Bates and Jackson, 1987).

Divalent ion—An atom that has acquired an electrical charge by gaining or losing two electrons.

Electric arc furnace—Steelmaking furnace in which scrap is generally 100 percent of the charge. Heat is supplied from electricity that arcs from the graphite electrodes to the metal bath. Furnaces may be either alternating current (AC) or direct current (DC). DC units consume less energy and fewer electrodes, but they are more expensive (*http://metals.about.com/library/bldef-Electric-Arc-Furnace.htm*).

Froth flotation—A method of mineral concentration used in platinum group metal production that separates the various minerals in the feed according to their differing surface properties. Separation is achieved by passing air bubbles through the mineral pulp. By adjusting the chemistry of the pulp with various reagents, valuable minerals can be made aerophilic (air avid) and gangue minerals aerophobic (water avid). Separation occurs when valuable minerals adhere to the air bubbles that form the froth floating on the surface of the pulp (available online at *http://www.platinum.matthey.com/production/1048863442.html*; accessed September 27, 2007).

Galvanized steel—Steel coated with a thin layer of zinc to provide corrosion resistance in underbody auto parts, garbage cans, storage tanks, or fencing wire. Sheet steel normally must be cold-rolled prior to the galvanizing stage (About Metals Glossary).

Grade—The relative quantity or percentage of the commodity or element of interest in a unit volume of mineralized rock.

Layered igneous complex (or layered intrusion)—A compositionally stratified intrusive magmatic rock body. Layering may be conspicuous because of variations in relative proportions of minerals.

Light-emitting diode (LED)—A semiconductor diode that converts applied voltage to light and is used in digital displays (e.g., a calculator). A diode is an electronic device that restricts current flow primarily to one direction (American Heritage Dictionary, 2000).

Liquid crystal display ("flat screen")—A low-power flat-panel display used in many laptop computers, calculators, and digital watches, made up of a liquid crystal that is sandwiched between layers of glass or plastic and becomes opaque when electric current passes through it. The contrast between the opaque and the transparent areas forms visible characters (American Heritage Dictionary, 2000).

Magnetic moment—A vector quantity characteristic of a magnetized body or an electric current system; it is proportional to the intensity of the magnetic field produced by this body and also to the force experienced in the magnetic field of another magnetized body or electric current (Bates and Jackson, 1987).

Metal—Any of a class of chemical elements that have a characteristic luster, are good conductors of heat and electricity, and are opaque, fusible, and generally malleable or ductile (Bates and Jackson, 1987).

Metallurgy—The science and art of separating metals and metallic minerals from their ores by mechanical and chemical processes; the preparation of more metalliferous materials from raw ore (Bates and Jackson, 1987).

Mineral deposit—A mineral occurrence of sufficient size and grade that it might, under favorable circumstances, be considered to have economic potential (available online at *http://www.coalcreek.org/glossary.html*; accessed November 19, 2007).

Mineral occurrence—Any concentration of ore or economic mineral found in bedrock or as float but that are too low grade or for other reasons are not considered potentially economic (USBM/USGS, 1980).

Mineral reserve—That part of the reserve base which could be economically extracted or produced at the time of determination with existing technology (USBM/USGS, 1980; Bates and Jackson, 1987).

Mineral reserve base—That part of an identified resource that meets specified minimum physical and chemical criteria related to current mining and production practices, including those for grade, quality, thickness, and depth. It is the in-place demonstrated resource from which reserves are estimated (USBM/USGS, 1980).

Mineral resource—A concentration of naturally occurring solid, liquid, or gaseous material in or on the Earth's crust in such form and amount that economic extraction of a commodity from the concentration is currently or potentially feasible. An "identified resource" is a resource whose location, grade, quality, and quantity are known or estimated from specific geologic evidence (USBM/USGS, 1980).

Mineralization—The process or processes by which a mineral or minerals are introduced into a rock, resulting in a valuable or potentially valuable deposit (Bates and Jackson, 1987).

Open-pit mine—Surface mining in which valuable rock is exposed by removal of overburden.

Ore—A naturally occurring material from which one or more minerals of economic value can be extracted at a reasonable profit (Bates and Jackson, 1987).

Ore body—A continuous, well-defined mass of material of sufficient ore content to make extraction economically feasible (Bates and Jackson, 1987).

Ore deposit—A mineral deposit of such grade, tonnage, or value that the minerals can be extracted, processed, and distributed at a profit.

Oxide—A mineral compound characterized by the linkage of oxygen with one or more metallic elements such as cuprite (Cu_2O), rutile (TiO_2), or spinel ($MgAl_2O_4$) (Bates and Jackson, 1987).

Platinum group metals—Ruthenium, rhodium, palladium, osmium, iridium, and platinum (Bates and Jackson, 1987).

Porphyry copper deposit—A large body of rock that contains disseminated chalcopyrite and other sulfide minerals. Such deposits are mined in bulk on a large scale, generally in open pits, for copper and sometimes by-product molybdenum. A porphyry is an igneous (magmatic) rock that contains conspicuous phenocrysts (mineral crystals) in a fine-grained groundmass (Bates and Jackson, 1987).

Pyrophoric alloy—An alloy that emits sparks when scratched or struck with steel (Webster's Third New International Dictionary, 1986).

Rare earth elements—A series of 15 metallic elements, from lanthanum (atomic number 57) to lutetium (71) that occur as oxides in the Earth's crust. These elements are not especially rare, but their concentrations are low. Three other elements (yttrium, thorium, and scandium) are sometimes also considered rare earth elements (Bates and Jackson, 1987).

Scanning electron microscope—A microscope in which a finely focused beam of electrons is electrically or magnetically moved across the specimen to be examined, from point to point, again and again, and the reflected and emitted electron intensity is measured and displayed, sequentially building up an image. The ultimate magnification and resolution is less than for the conventional electron microscope, but opaque objects can be examined and great depth of field is obtained (Bates and Jackson, 1987).

Semiconductor—A material that does not conduct electricity at low temperatures but does so at higher temperatures (Chemistry-Dictionary).

Strategic mineral—A mineral associated almost exclusively with national security and military needs or requirements during national emergencies; synonyms for strategic include planned, tactical, and calculated.

Superalloy—Any of several, complex, temperature-resistant alloys (American Heritage Dictionary, 2000).

Superconductor—A material that has no resistance to electricity. When current passes through a superconductor, there is no loss of electrical power due to these materials (Chemistry-Dictionary).

Sustainable resource development—The integration of economic activity with environmental integrity, social concerns, and effective governance systems (MMSD, 2002).

Titanium sponge—The metal product from reduction of titanium tetrachloride with magnesium (the Kroll process), called sponge because of its sponge-like appearance. Titanium sponge is highly susceptible to contamination and deterioration since its sponge-like characteristics permit pickup of free moisture if it is improperly packed and stored (available at *https://www.dnsc.dla.mil/iamthekey/UploadedFiles/GENERAL_CommodityData_titanium.pdf*; accessed September 27, 2007).

Tonnage—The amount of material containing the mineral commodity of interest.

Trivalent ion—An atom that has acquired an electric charge by gaining or losing three electrons.

References

About Metals Glossary. Available online at *http://metals.about.com/cs/governmentdata/l/blglossary_a.htm* (accessed September 26, 2007).

American Heritage Dictionary of the English Language, 2000. 4th ed. Boston: Houghton Mifflin. Available online at *http://www.bartleby.com/61/* (accessed September 27, 2007).

Bates, R., and J. Jackson (ed.), 1987. *Glossary of Geology*, 3rd ed. Alexandria, Va.: American Geological Institute.

Chemistry-Dictionary. Available online at *http://www.chemistry-dictionary.com/* (accessed September 26, 2007).

MMSD (Mining, Minerals, and Sustainable Development Project), 2002. *Breaking New Ground.* London: EarthScan Publications in association with the International Institute for Environment and Development, 441 pp.

Oxford Dictionary and Thesaurus, 1996. New York: Oxford University Press.

U.S. Bureau of Mines and U.S. Geological Survey (USBM/USGS), 1980. *Principles of a resource/reserve classification for minerals: U.S. Geological Survey Circular 831.* Reston, Va.: U.S. Geological Survey, 5 pp.

Webster's Third New International Dictionary, 1986. Springfield, Mass.: Merriam-Webster, Inc.

APPENDIX D

Periodic Table of Elements

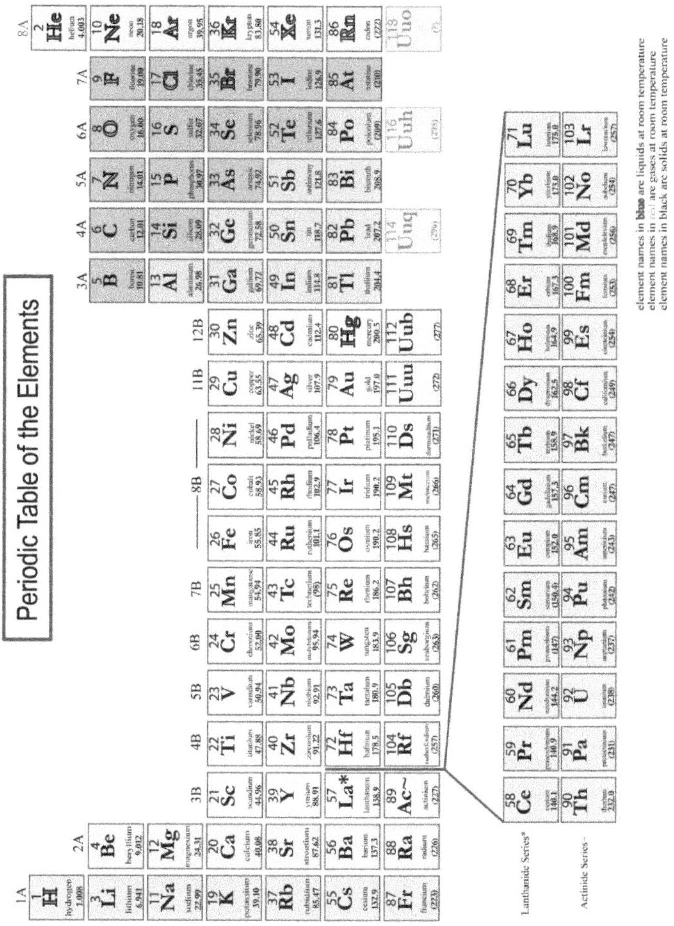

SOURCE: Los Alamos National Laboratory, Chemistry Division, *http://periodic.lanl.gov/downloads/periodictable.pdf.*